Drying and Valorisation of Food Processing Waste

Drying and Valorisation of Food Processing Waste is a comprehensive guide that delves into the crucial role of advanced drying technologies in mitigating the issue of food waste. This book evaluates the current research, technologies, and methodologies in food waste processing and valorisation, highlighting the challenges and opportunities that exist in this field.

This book provides a systematic classification of various types of food waste and how to choose the most appropriate drying technology based on waste characteristics. It also covers the effects of drying technologies on physical and chemical properties, as well as valuable compounds. In addition, it evaluates the impact of drying on different valorisation routes and provides real-life industrial case studies to illustrate the practical applications of the concepts discussed. It is an invaluable resource for professionals, researchers, and academics who are looking to gain a deeper understanding of the impact of drying on food waste reduction and valorisation.

This book is aimed at chemical, food, and environmental engineers as well as researchers and academics in these fields. It provides a comprehensive overview of the latest developments in food waste processing and valorisation and is an essential reference for professionals seeking to advance their knowledge in this field. Additionally, this book's practical approach and case studies make it an ideal resource for students and academicians who are looking to gain hands-on experience in food waste reduction and valorisation.

Advances in Drying Science and Technology

Series Editor

Arun S. Mujumdar
McGill University, Quebec, Canada

Drying and Roasting of Cocoa and Coffee
Ching Lik Hii and Flavio Meira Borem

Heat and Mass Transfer in Drying of Porous Media
Peng Xu, Agus P. Sasmito, and Arun S. Mujumdar

Freeze Drying of Pharmaceutical Products
Davide Fissore, Roberto Pisano, and Antonello Barresi

Frontiers in Spray Drying
Nan Fu, Jie Xiao, Meng Wai Woo, Xiao Dong Chen

Drying in the Dairy Industry
*Cécile Le Floch-Fouere, Pierre Schuck, Gaëlle Tanguy,
Luca Lanotte, Romain Jeantet*

Spray Drying Encapsulation of Bioactive Materials
Seid Mahdi Jafari and Ali Rashidinejad

Flame Spray Drying: Equipment, Mechanism, and Perspectives
Mariia Sobulska and Ireneusz Zbicinski

Advanced Micro-Level Experimental Techniques for Food Drying and Processing Applications
Azharul Karim, Sabrina Fawzia and Mohammad Mahbubur Rahman

Mass Transfer Driven Evaporation of Capillary Porous Media
Rui Wu, Marc Prat

Particulate Drying: Techniques and Industry Applications
Jangam Vinayak, Chung-Lim Law, and Shivanand Shirkole

Drying and Valorisation of Food Processing Waste
*Chien Hwa Chong, Rafeah Wahi, Chee Ming Choo, Shee Jia Chew, and
Mackingsley Kushan Dassanayake*

For more information about this series, please visit: www.routledge.com/Advances-in-Drying-Science-and-Technology/book-series/CRCADVSCITEC

Drying and Valorisation of Food Processing Waste

Chien Hwa Chong
Rafeah Wahi
Chee Ming Choo
Shee Jia Chew
Mackingsley Kushan Dassanayake

CRC Press
Taylor & Francis Group
Boca Raton London New York

CRC Press is an imprint of the
Taylor & Francis Group, an **informa** business

Cover image © Shutterstock

First edition published 2024
by CRC Press
2385 Executive Center Drive, Suite 320, Boca Raton FL 33431

and by CRC Press
4 Park Square, Milton Park, Abingdon, Oxon, OX14 4RN

CRC Press is an imprint of Taylor & Francis Group, LLC

© 2024 Chien Hwa Chong, Rafeah Wahi, Chee Ming Choo, Shee Jia Chew, and Mackingsley Kushan Dassanayake

ISBN: 978-1-032-32087-8 (hbk)
ISBN: 978-1-032-32088-5 (pbk)
ISBN: 978-1-003-31280-2 (ebk)

DOI: 10.1201/9781003312802

Typeset in Palatino
by codeMantra

Contents

Advances in Drying Science and Technology

Series Editor: Dr. Arun S. Mujumdar

It is well known that the unit operation of drying is a highly energy-intensive operation encountered in diverse industrial sectors ranging from agricultural processing, ceramics, chemicals, minerals processing, pulp and paper, pharmaceuticals, coal polymer, food, forest products industries as well as waste management. Drying also determines the quality of the final dried products. The need to make drying technologies sustainable and cost-effective via application of modern scientific techniques is the goal of academic as well as industrial R&D activities around the world.

Drying is a truly multi- and inter-disciplinary area. Over the last four decades, the scientific and technical literature on drying has seen exponential growth. The continuously rising interest in this field is also evident from the success of numerous international conferences devoted to drying science and technology.

The establishment of this new series of books entitled *Advances in Drying Science and Technology* is designed to provide authoritative and critical reviews and monographs focusing on current developments as well as future needs. It is expected that books in this series will be valuable to academic researchers as well as industry personnel involved in any aspect of drying and dewatering.

The series will also encompass themes and topics closely associated with drying operations, e.g., mechanical dewatering, energy savings in drying, environmental aspects, life cycle analysis, technoeconomics of drying, electrotechnologies, and control and safety aspects.

About the Series Editor

Dr. Arun S. Mujumdar is an internationally acclaimed expert in drying science and technologies. He is the Founding Chair in 1978 of the International Drying Symposium (IDS) series and Editor-in-Chief of *Drying Technology: An International Journal* since 1988. The fourth enhanced edition of his *Handbook of Industrial Drying* published by CRC Press has just appeared. He is recipient of numerous international awards including honorary doctorates from Lodz Technical University, Poland and University of Lyon, France.

Please visit www.arunmujumdar.com for further details.

Preface

The Food and Agriculture Organisation of the United Nations in 2023 estimated that total edible food wastage amounts to 1.3 billion tonnes and results in a carbon footprint of 3.3 billion tonnes of CO_2 equivalent to GHG released into the atmosphere annually. Thus, this book aims to provide information to readers regarding various drying technologies available for managing different types of food waste, and operating conditions for edible and non-edible final use. In Chapter 1, we discuss drying and valorisation as potential solutions for the utilisation of food processing waste. Emphasis was given to the UN's sustainable development goals which are directly linked with food insecurity as well as food waste management and highlight the role of drying in food waste management. Chapter 2 discusses food waste characterisation and valuable compounds. Following this, Chapter 3 discusses food waste drying characteristics, kinetics, and the different drying techniques to cater to different types of food waste. More details on the selection of drying technologies based on waste characteristics and valorisation routes are discussed in Chapter 4. In addition, this book presents examples and case studies of real industrial applications of food waste drying and valorisation.

Acknowledgement

The authors would like to express their deepest appreciation to Zhenjiang Zhinong Food Co. Ltd for providing materials and photos for Chapter 4.

Authors

Chien Hwa Chong is an Associate Professor at the Department of Chemical and Environmental Engineering, Faculty of Science and Engineering, University of Nottingham. He is an expert in the drying of biomaterials and wastewater treatment technology.
Google Scholar: https://scholar.google.com/citations?user=QgiR5KsA AAAJ&hl=en
Publons: https://publons.com/researcher/2464609/chien-hwa-chong/
ORCID: https://orcid.org/0000-0003-0877-1515

Rafeah Wahi is an Associate Professor in the Faculty of Resource Science and Technology, Universiti Malaysia Sarawak. Her research work involves Waste Utilisation Technology including microwave pyrolysis for conversion of sewage sludge and agricultural waste into value-added products. She is also involved in composting research, particularly on the agricultural waste, mushroom media residues, and food wastes.
Google scholar: https://scholar.google.com.my/citations?user=XDQF0e4AA AAJ&hl=en
Publon: https://publons.com/researcher/1903349/rafeah-wahi/
ORCID: https://orcid.org/0000-0002-3860-3566

Chee Ming Choo is a Senior Lecturer in Chemical Engineering Programme, Faculty of Engineering, Built Environment and Information Technology (FoEBEIT), SEGi University. Choo is currently serving as Chairman of the Laboratory Safety Committee of FoEBEIT, SEGi. Choo's expertise is related to water and wastewater research, contaminated land management, and sustainability research.
Google Scholar: https://scholar.google.com/citations?user=dhbRF_MAAAAJ&hl=e
ORCID: https://orcid.org/0000-0003-0254-9091

Shee Jia Chew is registered with the Board of Engineers Malaysia, IChemE (UK), and the Institute of Engineers Malaysia (IEM). She is now working as a Senior Chemical Engineer in a waste-to-energy company that specialises in pyrolysis technology in Singapore.
ORCID: https://orcid.org/0000-0002-2669-5985

Mackingsley Kushan Dassanayake is a Ph.D. candidate from the School of Pharmacy, University of Nottingham Malaysia. His area of expertise is related to antimicrobial drug discovery.
ORCID: https://orcid.org/0000-0002-3068-4349

1

Introduction to Drying and Valorisation of Food Processing Waste

1.1 Introduction

Impact of food waste: The problem of waste and loss of opportunity for valorisation in food products are highlighted by three facts, viz. 1/3 of food produced for human consumption globally, about 1.6 billion tons per year, is lost or wasted; the cost of food waste globally is estimated at around USD 2.6 trillion; and food waste accounts for 4.4 giga-tons (Gt) of CO_2 eq. per year, which represents 8% of global anthropogenic GHG emissions (Jain et al., 2018).

Food processors, governmental bodies, and consumers are taking action regarding food production patterns and consumption to decrease the generation of by-products and side-streams and increase their circularity by developing nutrient-rich products (Granato et al., 2022). According to Granato et al. (2022), it can be done by enhancing the nutritional composition, for example, fibres, anti-microbials, and anti-oxidants of products via a multi-disciplinary approach to align the sustainability issues, technology, legislation matters, and clinical evidence of functionality. For entrepreneurs or investors from small and medium industries that plan to process or convert edible food waste, clinical trials are required including identifying control groups, human intervention studies health benefits, via in-vitro testing and interactions with human cell line tests. They can look for more information from different authorities around the world like Food and Drug Administration – FDA, the European Food Safety Authority (EFSA), Europe, the Food Safety and Quality Division (FSQD) of the Ministry of Health (MOH), Malaysia, Food Safety Laws & Legislation Australian Capital Territory, Australia, National Medical Products Administration (NMPA), China, U.S. Food and Drug Administration, United States.

The term "food waste" appears in any discussion related to international food problems. It is always defined as any food and inedible parts of food, removed from the food supply chain that can be recovered or disposed of. The lost or wasted food must be reduced for greater food security.

DOI: 10.1201/9781003312802-1

1.2 Food Waste Contributing to Food Insecurity

Food waste significantly affects the environment and contributes to food insecurity. According to a report (FAO, 2013), cereal wastes like rice and wasted rice impacted the carbon, blue water, and arable land due to the emission of methane, meat waste as high as 67% from Latin America impacted the environment in terms of land occupation and carbon footprint, fruit waste emerges as a blue water hot spot in Asia, Latin America, and Europe, and vegetable waste constitutes high carbon footprint in Asia, Europe, South, and South East Asia. Figure 1.1 shows the food waste factors contributing to food insecurity. It started with water used to produce food. Huge amounts of water are used in food production and processing. Second, eliminate food waste at all points in the food production and supply chain (damaged or spoiled), in retail settings (physical flaws or overbuying), and at the consumer level (unused food or food that has expired as a result of overbuying). Further to this, poor storage and transportation also contributed to food waste due to poor harvesting, drying, threshing, solarisation, storage, and inadequate infrastructure. Most of the food waste is toxic in nature due to bacteria, viruses, parasites, or chemical substances. It consists of bacteria like

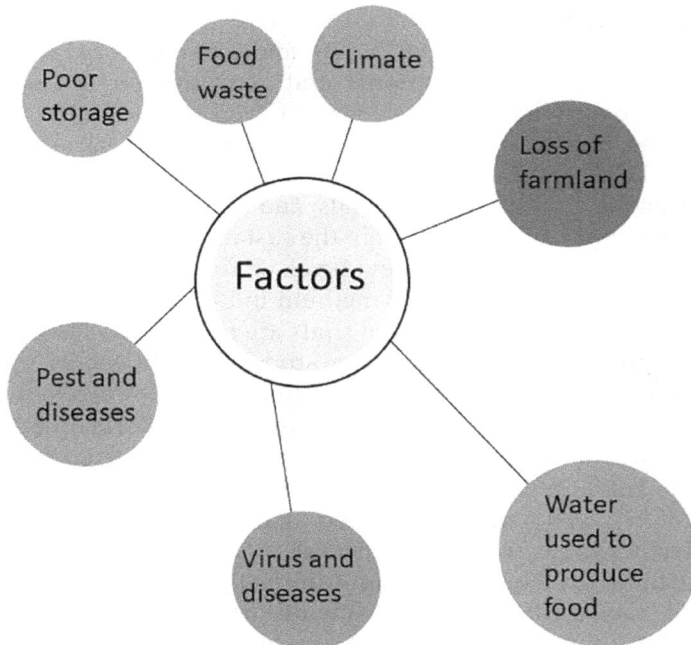

FIGURE 1.1
Food waste factors contributing to food insecurity.

Salmonella, Campylobacter, enterohaemorrhagic *Escherichia coli, Listeria,* and *Vibrio cholera;* viruses like norovirus and Hepatitis A virus; parasites like fish-borne trematodes, *Echinococcus* spp., *Taenia* spp., *Ascaris, Cryptosporidium, Entamoeba histolytica,* and *Giardia;* and Prions like Bovine spongiform enceph- alopathy (BSE) and Creduzfeldt-Jakob diseases (vCJD). All these factors lead to food insecurity and especially climate change.

Food waste has a huge impact on climate change. Climate change has a direct impact on food security as agricultural and fishery activities are heav- ily reliant on the climate. One of the major impacts of climate change is the increased frequency and intensity of climate-related disasters, such as floods, droughts, storms, and heat waves. According to a study carried out by World Meteorological Organisation (WMO), the number of extreme climate-related disasters has increased by a factor of five over the last 50 years. These disas- ters would cause physical damage to farms, crops, animals, and agricultural and food supply chain infrastructure, reducing agricultural productivity yields and food availability.

Agriculture is of fundamental importance to developing countries because a well-functioning agricultural sector is essential to ensuring food security, and agricultural products are a major source of national income. The main components of sustainable farming include soil management, water man- agement, disease/pest management, technology, and labour management. However, climate change has greatly affected agricultural activities in differ- ent aspects. For example, global warming would cause global sea water level rise and thus result in the permanent loss of coastal farmland, whereas the change of precipitation pattern will cause coastal flooding climate scenarios, resulting in the immediate loss of farm crops and agricultural infrastructure. Besides, climate change has resulted in rising global temperature and chang- ing patterns of precipitations, causing increased unpredictability in rainfall and streamflow, and decreased rainfall in many arid areas. As result, agri- cultural crops and livestock require more freshwater consumption to keep themselves hydrated.

1.3 Global Outlook on Food Waste

Food waste can be defined as food that is removed from the human food supply chain during the manufacturing, retail, and consumption stages, and it depends on consumer behaviour and government policies and regula- tion. Food waste can be divided into two categories 1 and 2. Category 1 is food waste from leftovers and trimmings from daily business due to fresh- ness issues. Food removed from the food supply chain can be recovered. It consists of unutilised edible resources in food production and unutilised raw materials which are not yet ready to be eaten. Category 2 is industry discards, where waste streams are generated in industrial food processing systems and agriculture residues from plant cultivation (crops ploughed in/not harvested). Edible food that has or had the potential to be eaten is removed from the food supply chain, and inedible parts of food are removed from the food supply chain. Food waste is a far-reaching issue, not only burdening the waste management system but also aggravating food security, the economy, and environmental sustainability. According to a study conducted by the Food and Agriculture Organisation of the United Nations in the year 2011, around 1.3 billion tons of food (one-third of the food produced) are lost or wasted annually. This amount of annual food waste is equivalent to more than the yearly production of cereal in the globe, according to the production data in the late 2000s. Both industrialised and developing countries contribute a nearly equal amount of food lost, where 670 and 630 million tons of food lost are contributed by industrialised and developing countries, respectively (FAO – News Article, n.d.). However, food loss in industrialised countries is dominant at retail and consumer levels, while in developing countries, more than 40% of food loss happens at the intermediate stages of the food supply chain, which are the post-harvest and processing stages.

In general, food is wasted at every stage of the food supply chain, from initial agricultural production to ultimate home consumption. Figure 1.2 shows food loss across food supply chain between middle- and high-income countries and Low income countries. Food is wasted in large amounts in middle- and high-income nations, which means that other than the food loss happening during the early stage in the food supply chain, food is discarded even though it is edible during the ultimate consumption stage. In contrast to distribution and consumption levels, these areas' agricultural production,

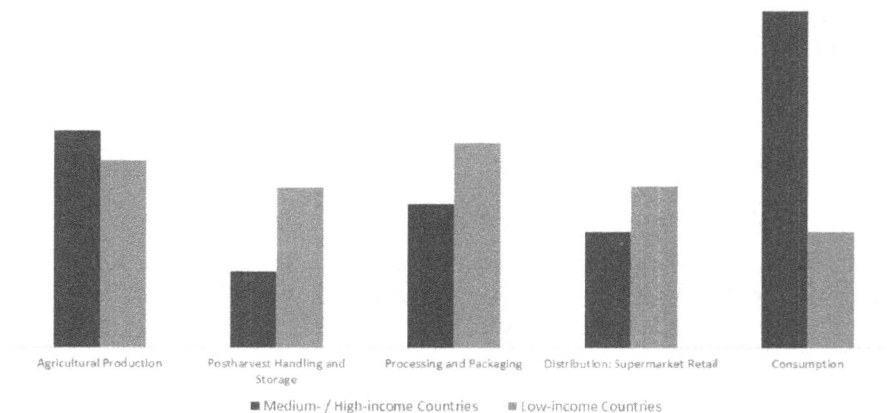

FIGURE 1.2
Food loss across food supply chain.

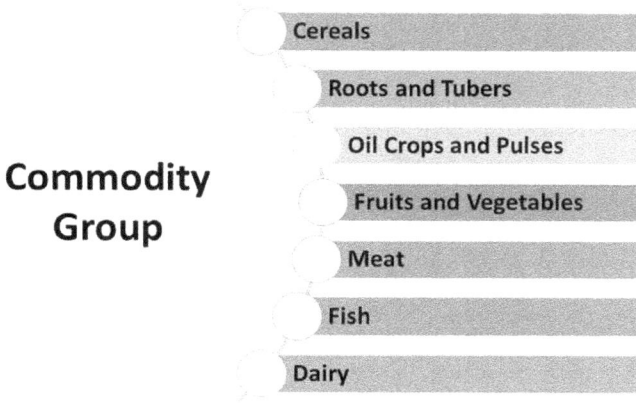

Cereals

Roots and Tubers

Oil Crops and Pulses

**Commodity
Group**

Fruits and Vegetables

Meat

Fish

Dairy

FIGURE 1.3
Categories of food produced based on commodity group. After Food and Agriculture
Organisation of the United Nation.

post-harvest processing, and storage are food supply chain phases with
particularly substantial food losses.

The Food and Agriculture Organisation of the United Nation has categorised the food produced into seven commodity groups (Figure 1.3), which
are cereals, roots and tubers, oil crops and pulses, fruits and vegetables,
meat, seafood, and dairy as presented in Table 1.1. The food loss across the
food supply chain for all commodity groups is studied and analysed across
the countries as shown in Tables 1.2 and 1.3. In the commodity group of
cereals, wheat is the primary crop supply in medium- and high-income
nations, and consumer losses account for around half of the total cereal
food waste. However, in the same commodity group, rice predominates as
a crop in low-income areas, particularly in the densely populated region of
South and Southeast Asia. Contrary to distribution and consumption levels,
these regions experience comparatively substantial food losses during agricultural production, post-harvest processing, and storage. The same situation also applies to other food commodity groups, where the major food
loss phase can be differed within the same group due to different food-
dominant regions.

To deal with this issue, United Nation Environment Programme (UNEP)
has proposed Sustainable Development Goal (SDG) 12.3, aiming to halve
the global food waste per capita at retail and consumer levels by year 2030.
Different actions have been taken by 193 countries to comply with SDG 12.3,
while many researchers have done some studies to reduce or recycle food
waste. Despite these multiple approaches, the generation of food waste cannot be avoided, and hence, the focus of the study has been shifted to the
valorisation of food waste.

TABLE 1.1

Commodity Groups Food Source and Food Waste (FAO, 2013)

No.	Commodity Groups	Food	Food Waste
1	Cereals (excluding beer)	1. Wheat 2. Rice (milled) 3. Barley 4. Maize 5. Rye 6. Oats 7. Millet 8. Sorghum 9. Other cereals	Edible Non-edible Food waste from restaurants and household Expired Contaminated
2	Root and tuber	1. Potatoes 2. Sweet potatoes 3. Cassava 4. Yams 5. Other roots	Food waste from markets, restaurants, and household Expired Contaminated
3	Oil seeds and pulses	1. Soybeans 2. Groundnuts (shelled) 3. Sunflower seeds 4. Rape and mustard seed 5. Cottonseed 6. Coconuts 7. Sesame seed 8. Palm kernels 9. Olives 10. Other oil crops	Skin, seeds, edible, leaves, trunks, multilayer shells Non-edible Food waste from restaurants, markets, and houses Expired Contaminated
4	Fruits and vegetables	1. Oranges and mandarins 2. Lemons and limes 3. Grapefruit 4. Other citruses 5. Bananas 6. Plantains 7. Apples excluding cider 8. Pineapples 9. Dates 10. Grapes 11. Other fruit 12. Tomatoes 13. Onions 14. Other vegetables	Skin, seeds, edible Non-edible Food waste from restaurants, markets, and houses Expired Contaminated
5	Meat	1. Bovine meat 2. Mutton 3. Pork 4. Poultry meat 5. Other meat 6. Offal	Skin, bone, fats, internal organs non-edible Food waste from restaurants, markets, and houses Expired Contaminated

(Continued)

TABLE 1.1 (*Continued*)

Commodity Groups Food Source and Food Waste (FAO, 2013)

No.	Commodity Groups	Food	Food Waste
6	Fish and seafood	1. Freshwater fish 2. Demersal fish 3. Pelagic fish 4. Other marine fish 5. Crustaceans 6. Cephalopods 7. Other aquatic products 8. Aquatic mammal meat 9. Aquatic plants	Scale, skin, bone, internal organs, non-edible, shell Food waste from restaurants, markets, and houses Expired Contaminated
7	Dairy products	1. Milk	Expired milk

TABLE 1.2

Grouping of World Regions (FAO, 2013)

Levels	Countries
Medium-/high-income regions	• Europe • USA • Canada • Oceania • Industrialised Asia
Low-income regions	• Sub-Saharan Africa • North Africa • West Asia • Central Asia • South Asia • Southeast Asia • Latin America

1.4 Valorisation of Food Waste

Valorisation is the process of improving the quality of food waste by converting them into renewable sources such as fuels, energy, chemicals, or materials via thermochemical or biotechnological processes. The management of food waste can be most sustainably achieved through techniques such as composting, fermentation, anaerobic digestion, gasification, and pyrolysis (Figure 1.4). These techniques can significantly contribute to material and energy supply, reduce ecological and environmental impact, and potentially create new employment opportunities. Among all the food waste valorisation techniques, composting is the most used technique due to its easy application and low investment cost and knowledge required. Table 1.4 explains different food waste valorisation techniques, while Table 1.5 analyses the pros and cons.

TABLE 1.3

Major Food Loss Phase of Commodity Groups (FAO, 2013)

Food Commodity Groups	Food-Dominant Regions	Major Food Loss Phase
Cereals (wheat)	Medium- and high-income countries	Consumer phase
Cereals (rice)	Low-income countries	Production and storage phase
Roots and tubers (sweet potato)	Medium- and high-income countries	Production and consumer phase
Roots and tubers (cassava)	Low-income countries	Production and storage phase
Oilseeds and pulses (sunflower seed, soybeans)	Medium- and high-income countries	Production phase
Oilseeds and pulses (groundnut)	Medium- and high-income countries	Production phase
Fruits and vegetables	Low-, medium-, and high-income countries	Production phase
Meat	Industrialised regions	Consumer phase
Fish and seafood	Industrialised regions	Production phase
Milk	Industrialised regions	Production phase

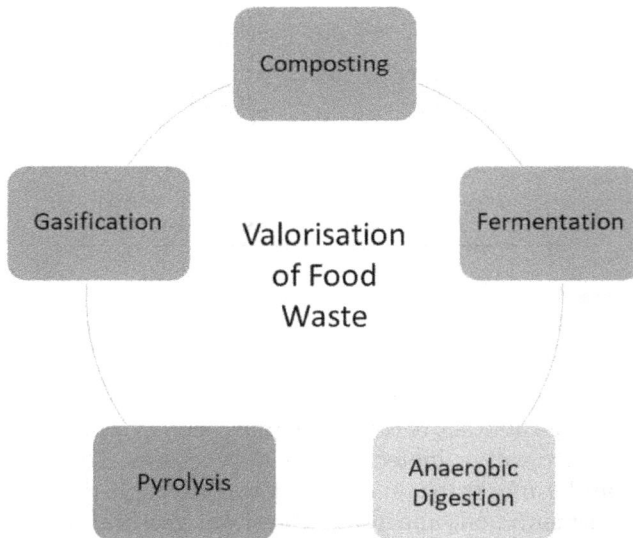

FIGURE 1.4

Management of food waste through various valorisation techniques.

TABLE 1.4

Food Waste Valorisation Techniques

Conversion Method	Explanation	References
Composting	• Process of breaking down organic material with the involvement of bacteria, oxygen, and water. • Relatively slow process (can take months). • Most common product is soil fertiliser.	EPA (n.d.-b) and Kazemi et al. (2017)
Fermentation	• Process involving the breakdown of complex organic nutrients into simpler carbon molecules with the assistance of micro-organisms under anaerobic or aerobic conditions. • One of the major products is ethanol.	Science Daily (n.d.) and Thiel et al. (1999)
Anaerobic digestion	• Conversion of organic matter into the production of methane abundant biogas in the absence of oxygen by bacteria.	Hong et al. (2013) and Meegoda et al. (2018)
Pyrolysis	• Thermal decomposition process in the absence of oxygen. • Consists of two variations, which are slow and fast pyrolysis. • Three primary products, which are syngas, bio-oil, and biochar.	Lee et al. (2019) and Meegoda et al. (2018)
Gasification	• Thermal decomposition process in the presence of small amounts of oxygen. • Primary product is syngas.	Lee et al. (2019)

1.5 Role of Drying in Food Waste Management

Drying is a valuable step in the valorisation of food waste either before, during, or after composting, fermentation, anaerobic digestion, gasification, and pyrolysis. It is known to reduce the mass and volume of food waste by as high as 90% using the commercial dryer. Further to this, drying technology is promoted as a potential food waste treatment technique to extract the value of food waste rather than just reducing the weight of the food waste. The reason behind this is that the physical and chemical properties of dried food waste have shown their potential to be converted into value-added products such as animal feeds and biofuels. The traditional drying technology used for food waste is sun and air drying. However, it is time-consuming, and the drying performance depends significantly on the weather. Therefore, different types of drying technologies have been developed to improve food waste drying efficiency. The technologies include thermal drying, bio-drying, combined solar drying, and microwave drying. In recent years, many studies have been carried out to investigate the impact of drying of food waste on the physical and chemical properties of food waste, the variables that affect the drying

TABLE 1.5

Pros and Cons of the Existing Valorisation Techniques

Conversion Method	Benefits	Drawbacks
Composting	Fruit and vegetable wastes consist of high levels of nutrients, nitrogen, phosphorous, potassium, calcium, magnesium, sodium, etc.	Higher transportation costs due to higher water content
	Can be applied as a soil amendment	Losses of nutrients that are composed of carbon, nitrogen and ammonia, phosphorous, potassium, and sodium as a result of leaching and microbial decomposition
	Increased carbon storage and water holding capacity	Composting is time-consuming (takes about 3–4months)
	Reduced greenhouse gas emissions	Potential microplastics contaminant within a substrate
	Sterile and stable and can be co-composted with other products to enhance the release of nutrients	Odour emissions
		Fruit and vegetable waste compost consist of lower amounts of trace elements
		Addition of excessive potassium levels to the environment
		Concentrated fruit and vegetable waste compost can release leachate that has high electrical conductivity and is acidic in nature
Anaerobic digestion	High availability of nitrogen, phosphorous, and potassium nutrients for uptake of plants which can be used as a soil amendment	Expensive transportation due to diluted nutrient content
	Reduced greenhouse gas emissions	Higher operational and capital cost of digestion plants establishment
	Lower odour emissions	Digestates may contain higher amounts of organic pollutants and heavy metals and food waste often gets mixed with sewerage sludge during the digestion process
	Reduced in pathogenic substance from the original substance	
	Increased storage of carbon and water holding capacity	
	Sterile and stable	

(Continued)

TABLE 1.5 (*Continued*)

Pros and Cons of the Existing Valorisation Techniques

Conversion Method	Benefits	Drawbacks
Pyrolysis	Increased potassium levels can be applied as a soil amendment	Higher operational and capital costs for machinery to produce biochar
	Higher rates of nitrification and ammonification when co-composted with food waste	Excessive release of potassium into the environment
	Reduced greenhouse gas emissions	
	Sequestration of carbon	
	Higher water holding and carbon storage capacity	
	Amending soil acids and increase of soil pH	
	Potential to remediate water and soil contaminants like organic pollutants and heavy metals	
	Sterile product	
	Lower cost	
Fermentation	Better mass and heat transfer can be achieved	Build-up of heat
	Greater diffusion of micro-organisms	Difficult to control process parameters and large-scale inoculums
	Commercially available on a large scale	Scale-up difficulties
Gasification	Less oxygen is used, and fewer emissions generated	Oxidation of nutrient compounds like sulphur and nitrogen

rate, challenges faced during drying of food waste, and the comparison of different drying technologies on food waste. However, a fundamental theory and applications of food waste using different drying technologies are rather scarce. Additionally, the influence of drying methods on physical and chemical properties and on other valuable compounds of food waste is limited. In addition, what are the industry directions and initiatives? Therefore, this monograph attempts to provide a detailed study of the influence of drying technologies, which has an impact on food waste's physical and chemical properties and other valuable compounds (detailed in Chapter 2).

Drying is one of the oldest technologies being utilised in food recovery systems, mainly being treated as one of the food preservation methods. It is used to preserve the food, prolong its shelf-life, and thus achieve the goal of minimising food waste. Sun hot air drying and bio-drying are the classic methods used for drying food waste. The drying process takes a long time, and the weather has a big impact on how well it works. As a result, many drying technologies, including combined drying, hybrid drying, and microwave drying, have been developed in recent years to increase drying efficiency.

Nowadays, it is promoted as a potential food waste treatment technique to retained chemical/bioactive compounds extract the value of food waste. The reason behind this is that the physical and chemical properties of dried food waste have shown their potential to be converted into value-added products such as animal feeds and biofuels (Schroeder et al., 2020). Besides, it is also found that drying technology can be a useful technique in the encapsulation of phenolic compounds extracted from food waste.

For the industries that pass their food waste to professional food waste handlers, high transportation costs and high emissions of greenhouse gases occurred due to the frequent waste collection required. This is because food waste is usually high in moisture content, and hence their mass and volume are high. This results in the large space required for storage and packaging. Since drying is effective in reducing the volume and mass of food waste, this issue can be avoided by pre-treating food waste using drying technology because the conventional food waste dryer can create a low-moisture environment, which is around 10% of the moisture level.

Fruit and vegetable waste is a potential material for animal feed due to its rich nutritional value. A study conducted by García et al. shows that fruit and vegetable waste consists of around 65% of nitrogen-free extract content on a dry basis, proving that it can be a potential energy source in animal diets (Esteban et al., 2007; Pathak et al., 2015). However, the high water content of around 80% in fruit and vegetable waste will accelerate microbial growth, resulting in waste-handling difficulty. Hence, it is recommended to dry the fruit and vegetable waste before proceeding to the next stage. Besides, moisture removal can also reduce the concentration of anti-nutritional factors, thus increasing the nutritional value of the animal feed. It is also recommended to cautiously utilise this technology as drying at high temperatures might alter the chemical nature and nutritional quality of fruit and vegetable waste, and thus drying at 65°C for 20 minutes is recommended for the waste to be dehydrated without losing the nutrition (García et al., 2005; San Martin et al., 2016).

Food waste containing animal by-products is a great potential source for biodiesel production due to the contents of oil and fats in food waste. It is always recommended to dehydrate the food waste before conversion into biodiesel. This is because the presence of water molecules in the food waste may inhibit the penetration of solvent into the food particles. However, it is advised to avoid the drying temperature exceeding 105°C, although a higher drying temperature can shorten the drying duration. This is because high drying temperatures might burn the food waste and break the fatty acid chain, affecting the quality of biodiesel produced (Barik et al., 2018) and extended the drying duration to 24 hours can lead to bone dry food waste.

Since the current industry used drying techniques by consuming around 20%–25% of the total energy used in the food processing industry, researchers have put their focus on more effective and economically friendly drying technology. Masud et al. (2020) studied the feasibility of utilising waste heat of flue gas generated by diesel engines, and the results showed that the drying performance and product quality conformed to the current industry standard. Besides, it is more environmentally sustainable compared to the electric convective drying techniques as waste heat is being recovered and emission of additional flue gas can be avoided (Masud et al., 2020).

The Environmental Protection Agency (EPA) of the United States has proposed a food recovery hierarchy to prevent and minimise food waste (Figure 1.5) (US EPA, n.d.-a). The most preferred option for food waste management is reducing food waste at source such as delivering less food to individual stores on daily basis. If the excess of food production over food

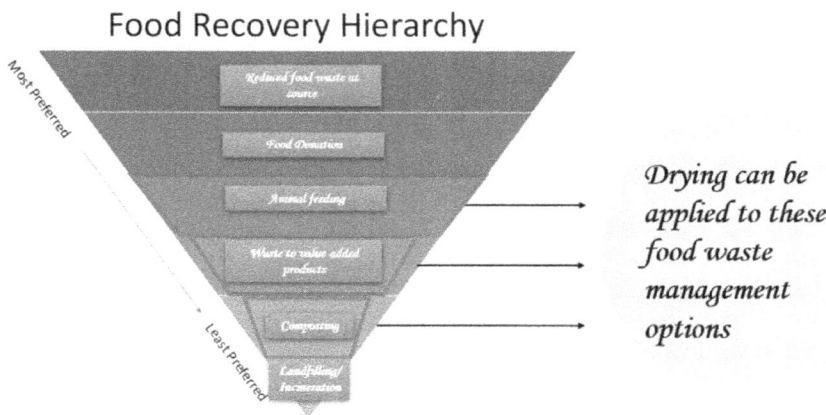

FIGURE 1.5
Food recovery hierarchy for prevention and minimisation of food waste. (Environmental Protection Agency (EPA) of United States).

consumption cannot be avoided, it is recommended to donate extra food to food banks and shelters. Although it is shown that landfill and incineration are the least preferred options for food waste handling, this technique is still the most used method as it is easier to be applied and relatively cost-effective compared to other disposal methods.

The practice of landfilling food waste is widely regarded as the least environmentally sustainable as it has brought several negative impacts to the environment (Table 1.6). First, landfilling of food waste contributes to around 20% of global methane emissions (Danthurebandara et al., 2012). The decomposition of food waste will generate methane, a greenhouse gas that can trap up to 20 times more heat than carbon dioxide, contributing to global warming (Trabold & Nair, 2018). Besides, the formation of highly toxic chemicals has occurred during the dissociation of food waste. As the rain falls, the rainwater rinses through those chemicals and flows into the groundwater, resulting in serious contamination of groundwater (Sciencing, n.d.).

Incineration of food waste is also one of the most widely used but least favourable options other than landfilling. Together with other municipal solid wastes, food waste is being fed into the incinerator for high-temperature combustion. Although it has been reported that 0.51 kg of carbon dioxide is released through the incineration of 1 kg of solid waste, the quality of combustion depends significantly on the characteristics of the waste fed into the incinerator (Wang & Geng, 2015). Unlike other combustion end products, the emissions of waste combustion usually carry ash, heavy metals, and other organic compounds, resulting in air pollution (Trabold & Nair, 2018).

Composting and anaerobic digestion are relatively preferable options compared to incineration and landfilling of food waste due to less impact on the environment. However, the emission of greenhouse gases still cannot be avoided through the application of the currently used technologies. Although there are many studies mentioning the potential environmental impact brought by conventional food waste management techniques, more research needs to be carried out to investigate the emission factor of each food waste management technique to ease the planning of a more effective management system, reducing the impact on the environment.

In recent years, many studies have been carried out to investigate the impact of drying of food waste on the physical and chemical properties of food waste, the variables that affect the drying rate, challenges faced during the drying of food waste, and the comparison of different drying technologies on food waste. Table 1.7 shows some selected publications of recent years' studies of drying technology on food waste, and the experimental results showed that drying technology can successfully reduce the mass and volume of food waste and slow down the biological decomposition of food waste. This can restrict odour emissions and increase the storage duration of food waste prior to the biological decomposition of food waste, easing the handling of it for the later steps of valorisation. Typically, the moisture reduction rate is depending significantly on the drying temperature. The higher the drying

TABLE 1.6

Role of Drying on Different Types of Food Waste Management Techniques

Types of Food Waste Management Techniques	Challenges	Role of Drying	References
Landfilling	Soil and groundwater pollution: dependent on the type of food waste landfilled, usually contains toxic elements, heavy metals, dissolved methane, fatty acids, sulphate, calcium, nitrate, phosphates, etc. Air pollution: emitting dust and greenhouse gases such as methane, carbon dioxide, etc.	• Reduce the mass and volume of food waste, thus reducing the transportation cost • Reduce the unpleasant odour	Abdullah et al. (2018), Iravanian and Ravari (2020), Tun and Juchelková (2018)
Incineration	Air pollution: pollutants generated depend on the components of food waste incinerated, usually carry ash, heavy metals, dioxins, and other organic and inorganic compounds	• Reduce process instability and loss of energy due to high moisture content in food waste	He et al. (2004), Shirinbakhsh et al. (2017), and Trabold and Nair (2018)
Composting	• Emission of unpleasant smell	• Reduce the mass and volume of food waste, thus reducing the transportation cost • Reduce unpleasant odour. Note: Food waste with moisture content that is lower than 50% might slow down the decomposition process	UGA Cooperative Extension (n.d.), Saer et al. (2013), and Schroeder et al. (2020)
Anaerobic digestion (biogas generation)	Non-methane volatile organic compounds (VOC) such as esters, ketones, and aromatic hydrocarbons, etc., are generated along the anaerobic digestion process	After the generation of biogas, digestate can be dried up and sold as fertiliser	

TABLE 1.7

Selected Findings Reported in the Literature Related to Laboratory-Scale Drying of Food Waste from 2016 to 2020

Types of Food Waste	Methods	Major Findings	References
Household	Thermal drying at a temperature of 80°C for 8 hours	• Mass reduction of about 80%. • Volume reduction of about 60%. • Longer storage time since biodegradation is avoided • Reduced sugar is preserved	Sotiropoulos et al. (2016)
University canteen	Thermally assisted bio-drying	• Lag phase is avoided during the heating acclimation stage • Around 13% of thermal energy is used to trigger bio-drying; thus, the remaining thermal energy is used for water removal • Water removal rate was promoted significantly with increasing air flow rates (the quality of dried products with a thermal assisted bio-drying air flow rate of 0.8 L/min/kg after 4 days is the same as the products dried with conventional bio-drying after 20 days)	Ma et al. (2018)
Restaurant	Thermal drying	• Around 50% of heat loss occurs during heat exchange between hot steam and air • The lower the dry air temperature, the longer the drying duration • It is recommended for direct steam drying for higher efficiency	Song et al. (2019)
Treatment facilities	Thermal drying	• It took around 8 hours to dry food waste of 100 kg, removing moisture to 18% • Condensate generated can be used as a deodorant and liquid fertiliser • Lower heating value of dried food waste is 4,431 kcal/kg, indicating it is a potential source of fuel • Odour of food waste can be prevented through drying of food waste	Han (2016)
Brewers' spent grain	Convective drying	• The higher the drying temperature and airflow, the shorter the drying time required	Arranz et al. (2018)
Coconut husk and shells	Sun drying Thermal drying	• Average calorific value of oven-drying coconut waste is 42% higher than that sun-drying coconut waste • Thermal drying coconut wastes emitted less carbon monoxide than sun-drying coconut wastes	Obeng et al. (2020)

(Continued)

TABLE 1.7 (*Continued*)

Selected Findings Reported in the Literature Related to Laboratory-Scale Drying of Food Waste from 2016 to 2020

Types of Food Waste	Methods	Major Findings	References
Oil palm frond waste	Thermal drying	• The higher the drying temperature, the higher the drying rate • The size of waste is shrunk by around 65%, and the shrinkage occurs in a radial direction • Drying kinetics is identified	Halim and Triwibowo (2016)
Grain corn	Thermal drying	• Instant cooling after high-temperature drying will lead to stress cracks in the kernel, resulting in value losing • Drawback of flat-bed dryer: high labour cost incurred for loading and unloading of food waste	Shukri et al. (2020)
Pre-consumed and post-consumed	Thermal drying	• The nutritional value of all types of dried food waste is lower than the nutritional value contained in conventional fertiliser • Dried food waste is not recommended for composting due to the emission of ammonia • Dried food waste has the potential to be soil amendment after being converted into biochar • Most of the dried food waste meet the nutritional requirement of fish feed; thus, it is recommended to be utilised as fish feed • It is recommended for dried food waste to be pelletised fuel due to its high energy content (average of 20 MJ/kg)	Schroeder et al. (2020)
Preserved food processing	Solar drying	• Air flow rate is inversely proportional to the air temperature, and thus, it is recommended to manipulate the air flow to maintain the drying temperature • Comparison between solar drying and sun drying: Moisture reduction rate of sun drying is lower than solar drying	Tony and Tayeb (2011)
Corn silk	Freeze drying Spray drying Microwave drying	• Most effective drying method, providing the highest value of phenolic compound • Best encapsulation of anti-oxidant phenolic compounds • Efficiency: freeze drying, spray drying, microwave drying • Bulk density of powder produced by microwave drying is the highest, followed by spray drying and freeze drying	Pashazadeh et al. (2021)

temperature, the shorter the drying period required to achieve the desired moisture content. Instant cooling of food waste after high-temperature drying will result in a value loss of food waste due to stress cracks. Therefore, it is recommended to manipulate the air flow rate to achieve the desired drying temperature. Moreover, a comparison of different drying technologies on food waste moisture removal rate is also determined by the scientists. It is shown that thermally assisted bio-drying of food waste with a drying air flow rate of 0.4 L/min can shorten the drying duration by 16 days compared to the conventional bio-drying technology used. Besides, thermal drying and solar drying of food waste show higher drying efficiency than traditionally used sun-drying technology (Obeng et al., 2020; Tony & Tayeb, 2011).

Hojjat Pashazadeh et al. have carried out a study to investigate the performance of different drying methods in the encapsulation of phenolic compounds extracted from corn silk. The result showed that freeze drying is most effective in moisture removal and encapsulation of anti-oxidant phenolic compounds, followed by spray drying and microwave drying. The bulk density of the powders produced is one variable studied as powders with low bulk densities are more likely to oxidise and thus result in lower storage stability. It is shown that the powder produced by microwave drying is the highest, followed by spray drying and freeze drying (Pashazadeh et al., 2021).

The following chapter will discuss the drying characteristics and kinetics of food processing waste and the mechanism of drying and waste reduction, effective diffusivities, and different types of drying technologies. Further to this, unsolved problems and detained explanations of some research papers will be presented as well.

References

Abdullah, M., Rosmadi, H. A., Azman, N. Q. M. K., Sebera, Q. U., Puteh, M. H., Muhamad, A., & Zaiton, S. N. A. (2018). Effective drying method in the utilization of food waste into compost materials using effective microbe (EM). *AIP Conference Proceedings*, 020120. https://doi.org/10.1063/1.5066761

Arranz, J. I., Miranda, M. T., Sepúlveda, F. J., Montero, I., & Rojas, C. V. (2018). Analysis of Drying of Brewers' Spent Grain. *The 2nd International Research Conference on Sustainable Energy, Engineering, Materials and Environment*, 1467. https://doi.org/10.3390/proceedings2231467

Barik, S., Paul, K., & Priyadarshi, D. (2018). Utilization of kitchen food waste for biodiesel production. *IOP Conference Series: Earth and Environmental Science, 167*, 012036. https://doi.org/10.1088/1755-1315/167/1/012036

Danthurebandara, M., van Passel, S., , Nelen, D., Tielemans, Y., & van Acker, K. (2012). Environmental and Socio-economic Impacts of Landfills. Linnaeus ECO-TECH 2012, Kalmar, Sweden, November 26–28, 40–52.

Esteban, M. B., García, A. J., Ramos, P., & Márquez, M. C. (2007). Evaluation of fruit–vegetable and fish wastes as alternative feedstuffs in pig diets. *Waste Management*, 27(2), 193–200. https://doi.org/10.1016/j.wasman.2006.01.004

FAO. (2013). *Food wastage footprint*. https://www.fao.org/nr/sustainability/food-loss-and-waste/en/

FAO – News Article. *Cutting Food Waste to Feed the World*. (n.d.). Retrieved January 11, 2023, from https://www.fao.org/news/story/en/item/74192/icode/

García, A. J., Esteban, M. B., Márquez, M. C., & Ramos, P. (2005). Biodegradable municipal solid waste: characterization and potential use as animal feedstuffs. *Waste Management*, 25(8), 780–787. https://doi.org/10.1016/j.wasman.2005.01.006

Granato, D., Carocho, M., Barros, L., Zabetakis, I., Mocan, A., Tsoupras, A., Cruz, A. G., & Pimentel, T. C. (2022). Implementation of sustainable development goals in the dairy sector: Perspectives on the use of agro-industrial side-streams to design functional foods. *Trends in Food Science and Technology*, 124(April), 128–139. https://doi.org/10.1016/j.tifs.2022.04.009

Halim, A., & Triwibowo, B. (2016). *Drying Kinetics of Oil Palm Frond Waste Using Simple Batch Oven Dryer*. https://doi.org/10.15294/jbat.v4i2.4151

Han, D. H. (2016). A recycling method of food waste by drying and fuelizing. *Journal of Ecophysiology and Occupational Health*, 16(1–2), 52–58. https://doi.org/10.15512/joeoh/2016/v16i1&2/15637

He, P., Zhang, H., Zhang, C., & Lee, D. (2004). Characteristics of air pollution control residues of MSW incineration plant in Shanghai. *Journal of Hazardous Materials*, 116(3), 229–237. https://doi.org/10.1016/j.jhazmat.2004.09.009

Hong, Y., Nizami, A.-S., Pour Bafrani, M., Saville, B. A., & MacLean, H. L. (2013). Impact of cellulase production on environmental and financial metrics for ligno-cellulosic ethanol. *Biofuels, Bioproducts and Biorefining*, 7(3), 303–313. https://doi.org/10.1002/bbb.1393

Iravanian, A., & Ravari, S. O. (2020). Types of contamination in landfills and effects on the environment: A review study. *IOP Conference Series: Earth and Environmental Science*, 614(1), 012083. https://doi.org/10.1088/1755-1315/614/1/012083

Jain, S., Newman, D., Cepeda-Márquez, R., & Zeller, K. (2018). Global food waste management: Full report an implementation guide for cities. World Biogas Association, 1–145. http://www.waste.ccacoalition.org/document/white-paper-waste-and-climate-change-iswa-key-

Kazemi, K., Zhang, B., & Lye, L. M. (2017). Assessment of microbial communities and their relationship with enzymatic activity during composting. *World Journal of Engineering and Technology*, 05(03), 93–102. https://doi.org/10.4236/wjet.2017.53B011

Lee, S. Y., Sankaran, R., Chew, K. W., Tan, C. H., Krishnamoorthy, R., Chu, D.-T., & Show, P.-L. (2019). Waste to bioenergy: A review on the recent conversion technologies. *BMC Energy*, 1(1). https://doi.org/10.1186/s42500-019-0004-7

Ma, J., Zhang, L., Mu, L., Zhu, K., & Li, A. (2018). Thermally assisted bio-drying of food waste: synergistic enhancement and energetic evaluation. *Waste Management*, 80, 327–338. https://doi.org/10.1016/j.wasman.2018.09.023

Masud, M. H., Ananno, A. A., Ahmed, N., Dabnichki, P., & Salehin, K. N. (2020). Experimental investigation of a novel waste heat based food drying system. *Journal of Food Engineering*, 281, 110002. https://doi.org/10.1016/j.jfoodeng.2020.110002

Meegoda, J., Li, B., Patel, K., & Wang, L. (2018). A review of the processes, parameters, and optimization of anaerobic digestion. *International Journal of Environmental Research and Public Health, 15*(10), 2224. https://doi.org/10.3390/ijerph15102224

Obeng, G. Y., Amoah, D. Y., Opoku, R., Sekyere, C. K. K., Adjei, E. A., & Mensah, E. (2020). Coconut wastes as bioresource for sustainable energy: quantifying wastes, calorific values and emissions in Ghana. *Energies, 13*(9), 2178. https://doi.org/10.3390/en13092178

Pashazadeh, H., Zannou, O., Ghellam, M., Koca, I., Galanakis, C. M., & Aldawoud, T. M. S. (2021). Optimization and encapsulation of phenolic compounds extracted from maize waste by freeze-drying, spray-drying, and microwave-drying using maltodextrin. *Foods, 10*(6), 1396. https://doi.org/10.3390/foods10061396

Pathak, P. D., Mandavgane, S. A., & Kulkarni, B. D. (2015). Fruit peel waste as a novel low-cost bio adsorbent. *Reviews in Chemical Engineering, 31*(4). https://doi.org/10.1515/revce-2014-0041

Saer, A., Lansing, S., Davitt, N. H., & Graves, R. E. (2013). Life cycle assessment of a food waste composting system: Environmental impact hotspots. *Journal of Cleaner Production, 52*, 234–244. https://doi.org/10.1016/j.jclepro.2013.03.022

San Martin, D., Ramos, S., & Zufía, J. (2016). Valorisation of food waste to produce new raw materials for animal feed. *Food Chemistry, 198*, 68–74. https://doi.org/10.1016/j.foodchem.2015.11.035

Schroeder, J. T., Labuzetta, A. L., & Trabold, T. A. (2020). Assessment of dehydration as a commercial-scale food waste valorization strategy. *Sustainability, 12*(15), 5959. https://doi.org/10.3390/su12155959

Shirinbakhsh, M., Amidpour, M., & Nasir, K. (2017). Design and optimization of solar-assisted conveyer-belt dryer for biomass. *Energy Equipment and Systems, 5*(2), 1–10. http://energyequipsys.ut.ac.irwww.energyequipsys.com

Shukri, J., Syariffuddeen, A., & Redzuan, A. (2020). Evaluation on drying temperature of grain corn and its quality using flat-bed dryer. *ASM Science Journal, 13*(4), 78–83. https://www.akademisains.gov.my/asmsj/article/evaluation-on-drying-temperature-of-grain-corn-and-its-quality-using-flat-bed-dryer/

Song, D. B., Lim, K. H., & Jung, D. H. (2019). Developing heated air dryer for food waste using steam generated from incineration plant. *Journal of Biosystems Engineering, 44*(2), 112–119. https://doi.org/10.1007/s42853-019-00021-1

Sotiropoulos, A., Bava, N., Valta, K., Vakalis, S., Panaretou, V., Novacovic, J., & Malamis, D. (2016). Household food waste dehydration technique as a pretreatment method for food waste minimisation. *International Journal of Environment and Waste Management, 17*(3/4), 273. https://doi.org/10.1504/IJEWM.2016.078598

Sciencing. *The Effects of Landfills on the Environment.* (n.d.). Retrieved January 12, 2023, from https://sciencing.com/effects-landfills-environment-8662463.html

Thiel, T., Cohn, D., Kalk, M. R., Bruce Hemming, I. C., Reynolds, R., May, V. L., Alters, S., Alters, B., Pringle, A., & Corbin, D. (1999). *Science in the Real World Microbes in Action.* Kimber Mallet.

Tony, M. A., & Tayeb, A. M. (2011). *The Use of Solar Energy in a Low-Cost Drying System for Solid Waste Management: Concept, Design and Performance Analysis*, Euraasia Waste Management Symposium, November 14–16- Haliç Congress Center Istanbul/Turkey.

Trabold, T. A., & Nair, V. (2018). Conventional Food Waste Management Methods. In Trabold, T.A., Babbitt, C.W. (Eds.), *Sustainable Food Waste-To-Energy Systems* (pp. 29–45). Elsevier. https://doi.org/10.1016/B978-0-12-811157-4.00003-6

Tun, M. M., & Juchelková, D. (2018). Drying methods for municipal solid waste quality improvement in the developed and developing countries: A review. *Environmental Engineering Research*, 24(4), 529–542. https://doi.org/10.4491/eer.2018.327

Science Daily. *Turning Food Waste Back into Food: Fermenting Used Food Can Improve Crop Growth*. (n.d.). Retrieved January 11, 2023, from https://www.sciencedaily.com/releases/2021/01/210128091143.htm

UGA Cooperative Extension. *Food Waste Composting: Institutional and Industrial Application*. (n.d.). Retrieved January 12, 2023, from https://extension.uga.edu/publications/detail.html?number=B1189&title=food-waste-composting-institutional-and-industrial-application

US EPA. *Food Recovery Hierarchy*. (n.d.-a). Retrieved January 12, 2023, from https://www.epa.gov/sustainable-management-food/food-recovery-hierarchy

US EPA. *Reducing the Impact of Wasted Food by Feeding the Soil and Composting* (n.d.-b). Retrieved January 11, 2023, from https://www.epa.gov/sustainable-management-food/reducing-impact-wasted-food-feeding-soil-and-composting

Wang, Z., & Geng, L. (2015). Carbon emissions calculation from municipal solid waste and the influencing factors analysis in China. *Journal of Cleaner Production*, 104, 177–184. https://doi.org/10.1016/j.jclepro.2015.05.062

2

Food Waste Characteristics and Valuable Compounds

2.1 Food Waste Characteristics

Food waste can be characterised by its moisture/solid content, biodegradability, pH, variation in quality, electrical conductivity, and density. According to Selvam et al. (2021), food waste has the moisture content ranging from 48% to 95% and volatile solids ranging from 73% to 98%. For carbon, hydrogen, nitrogen, oxygen and sulphur (CHNOS) analysis, the percentage is 48%, 7%, 6.8%, 35%, 3.0%, and 0.3%, respectively. The other micro-nutrients are Na, Mg, and Ca with an average of 13.5, 1.1, and 11.3 g/kg. Carbohydrate, protein, and lipid contents ranged from 17% to 75%, 6% to 42%, and 5% to 44%, respectively (Selvam et al., 2021). Figure 2.1 shows the different types of fruit types food waste such as (a) orange peels, (b) litchi peels and seeds, (c) longan peels and seeds, (d) rambutan peels, (e) rose apple peels and flesh, (f) guava skins and flesh, (g) mango peels, (h) pineapple skins, (i) banana peels, and (j) dragon fruit peels. The shell and bone types food wastes such as (a) clam shell, (b) crab shells, (c) eggshells, (d) quail eggshells, (e) shrimp shells, (f) stomatopod shells, (g) chicken bones, and (h) fish bones are shown in Figure 2.2. Figure 2.3 shows vegetable, root, and tuber types of food wastes such as lotus root skins, corn silks, and skins, and Japanese sweet potato skins.

2.1.1 Moisture Content

Food waste moisture content must be measured before being subjected to valorisation processes such as composting, fermentation, anaerobic digestion, gasification, and pyrolysis. The moisture content of food waste has various significant dependants on the storage and characteristics of food waste due to the presence of heavy organic matter and thickness. The percentage of moisture content of food waste can be determined using a moisture analyser. If a moisture analyser is not available, an oven is required to determine the moisture content. It can be done by weighing the initial samples before drying using a convective hot air oven at desired operating conditions.

DOI: 10.1201/9781003312802-2

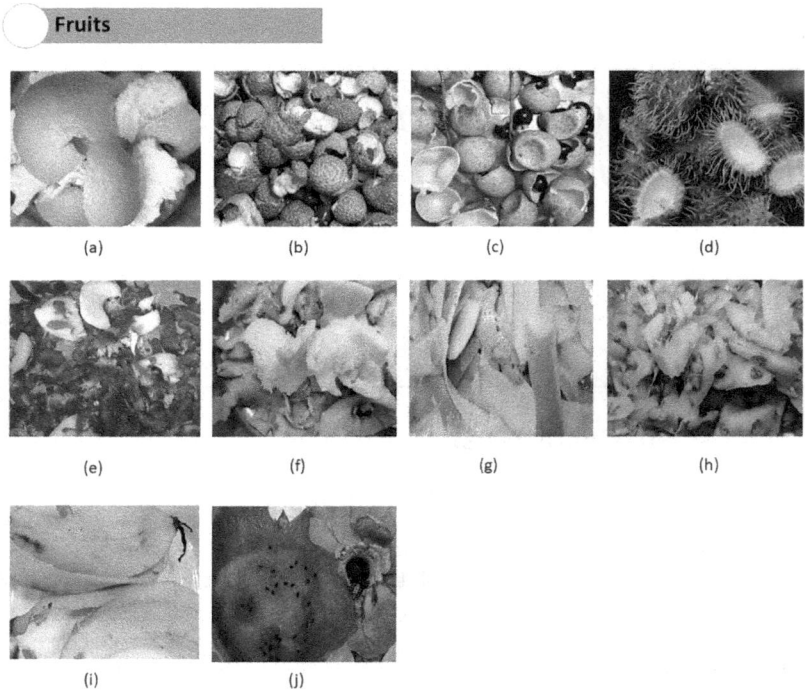

FIGURE 2.1
Different types of fruits types of food waste. (a) orange peels, (b) litchi peels and seeds, (c) longan peels and seeds, (d) rambutan peels, (e) rose apple peels and flesh, (f) guava skins and flesh, (g) mango peels, (h) pineapple skins, (i) banana peels, and (j) dragon fruit peels.

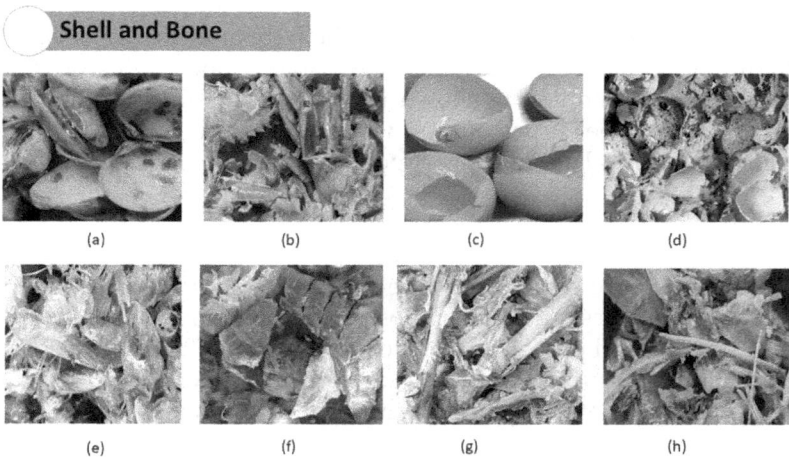

FIGURE 2.2
Shell and bone types of food wastes. (a) clam shell, (b) crab shells, (c) eggshells, (d) quail eggshells, (e) shrimp shells, (f) stomatopod shells, (g) chicken bones, and (h) fish bones.

Vegetables & Roots & Tubers

(a)　　　　　　　　(b)　　　　　　　　(c)

FIGURE 2.3
Vegetable, root, and tuber types of food wastes like (a) lotus root skins, (b) corn silks, and skins, and (c) Japanese sweet potato skins.

The samples can be dried until constant weight to determine the equilibrium moisture content. In the end, the samples need to be dried at 105°C for 24 hours using the AOAC method to obtain the bone dry weight. It can be determined using equation 2.1 (wet basis) or 2.2 (dry basis). It can also be analysed using a commercial moisture analyser.

$$MC(\%) = \frac{\left(W_i - W_f\right)}{W_i} \times 100\% \tag{2.1}$$

$$MC(\%) = \frac{\left(W_i - W_f\right)}{W_f} \times 100\% \tag{2.2}$$

where MC is the moisture content, W_i is the initial weight, and W_f is the final weight.

2.1.2 Biodegradability

Food waste has emerged as a viable waste stream due to its large volume and composition. However, some issues need to be resolved to successfully boost the global economy. These issues include seasonal variations in both quantity and composition, uncertainties in pre-treatment techniques, and storage issues.

Aerobic biodegradation may help solve some of these problems by acting as a pre-treatment method to hydrolyse food waste while providing liquefied wastewater. Many of these bio-based economic concepts rely on the use of microbes to convert hydrolysed food waste into usable chemicals, which can be converted into extractable carbohydrates, proteins, lipids, and secondary metabolites. Pre-treatment is an important step for conversion into the desired chemical constituents (Lin et al., 2013). The optimum pH for food waste-converting microorganisms such as methanogens and acidogens has been between 5.5–6.5 and 4.5–5.5, respectively (Wu et al., 2016).

2.1.3 pH

Different food wastes have different pH values. This property value will affect the microbes. According to Khalida et al. (2022), drying food waste will reduce the acidity of food waste and a typical pH ranged from 4.7 to 5.1. Fruit-vegetable waste, which comprises vegetable and fruit peelings, is high in lignocellulose waste. Lignocellulosic materials are more resistant to microbial degradation, which can cause acidification to slow down. Suitable drying methods can change the acidification of food waste. The release of ammonia or ammonium can improve the system via their buffering with the production of metabolic acids or emerge as suppressive to the process. The typical pH of food waste peeled from fruits ranged from 7 to 9. The swift degradation of volatile compounds of food waste can result in rapid acidification and accumulation of volatile fatty acids. Low pH can inhibit the activity of methanogens. The optimum pH for methanogens and acidogens has been between 5.5–6.5 and 4.5–5.5, respectively. The anaerobic digestion of food waste through a lower organic rate of loading was limited due to the high biodegradability of the organic matter.

2.1.4 Total Solids

The quality of food waste can be summarised based on its total solids (TS) and volatile solids. It has an impact on reactor size, the energy requirement for heating, water usage, and inactivation of pathogens during anaerobic digestion (Arelli et al., 2018; Jiang et al., 2018). The dry matter (DM) or TS of canteen food waste ranged from 16.7% to 24.87%, source-segregated food waste (SSFW) ranged from 23.7% to 25.9%, cafeteria food waste ranged from 18.9% to 22.6%, dining hall 13.68% to 23.1%, kitchen food waste 33.1%, and university restaurant food waste 23.2% to 25.1% (Selvam et al., 2021). The DM can be adjusted using drying technology.

2.1.5 Volatile Solids

Volatile matter (VS) or organic matter is a measure of the available organic content, the energy source for microbes in food waste treatment technologies. These materials that burn in a gaseous state are a fundamental tool for biomass energy calculations. High-volatile matters are good in utilisation of composting, anaerobic digestion, and other biological technologies (Selvam et al., 2021). Drying technologies plays a very important role here as it can reduce the moisture content to prevent the loss of saprophytic values during incineration. The drying method selected must be able to retain the volatile matter.

2.1.6 Electrical Conductivity

Electrical conductivity has an influence on the salinity of food waste. According to Aqeela et al. (2021), the pH value increased, while the electrical conductivity (EC) decreased during the food waste composting process.

The usage of salt during cooking leads to the high salt content in feedstock and consequently has a high EC value. Other studies also reported that food waste had a high EC value due to drying (Khalida et al., 2022). EC values of 8.9 m S/cm of food waste reported by Shi et al. (2016), 7.8 m S/cm (Donahue et al., 1998), and 5.1 m S/cm (Agapios et al., 2020) have a high electrical conductivity (EC) value (4.83–7.64 m S/cm) (Khalida et al., 2022).

2.1.7 Density

The density of food waste is calculated according to the following equation (2.3):

$$\text{Density}\,(kg/m^3) = \text{Weight}\,(kg)/\text{Volume}\,(m^3) \tag{2.3}$$

The density of the food waste generated ranges between 0.035 and 0.077 kg/m^3 for Café A and Café B within 15 days (Kamaruddin et al., 2020).

2.2 Food Waste Composition

The composition of food waste can be determined based on its physical and chemical composition. Most raw materials present in food waste comprise rice, bakery products, meat, fat, bones, fruit, and vegetables. Analysis of biodegradable food waste is selective for anaerobic digestion, pyrolysis, or transesterification. The typical composition of food waste is bakery products, rice, meat, fat, vegetables, fruits, and bone with 19%, 39%, 25%, 13%, 2% and 2%, respectively. It consists of 39% of water, 26% of carbohydrates, 15% crude fats, 17% crude protein, 0.3% fibre, and 3% ash.

2.3 Mechanical Properties of Food Waste

The data of various foods' modulus of elasticity have been collected to determine ultimate forces. The variants are from soft solids like vegetables to the hardest possible materials like animal bones. Table 2.1 shows the values of the specific limits of forces required for various types of foods.

2.4 Other Properties of Food Waste

The study of food waste characteristics is important in the food waste utilisation industry. For instance, it gives details about the micro-structure of a

TABLE 2.1

Mechanical Properties of Food Waste (Kim et al., 2019; Razali et al., 2013; Verma et al., 2020)

Type	Food Waste Material	Modulus of Elasticity (MPa)
Edible	Corn Kernels	0.000284
	Onions	219
	Apple	0.0014
	Wheat	5.2
	Carrot	239
	Bones	275.8
Non-edible	Meat	3.15
	Fish	0.005

food waste product. Products produced from food waste rely on the rate and nature of deformation that materials undergo when subjected to processing conditions. These parameters can be applied to predict product quality and determine the energy required. It helps in determining the quality of food waste products, design, and selection of food waste dryers. Below are some application examples.

a. Combining two or more food wastes that are manually or mechanically blended.

b. Flow rate control of food waste during processing. The flowability varies from low to highly viscous food waste.

c. Dispensing of food waste. Food waste can come out smoothly or roughly after processing.

d. Floating or settling of food waste with variable d specific gravity can either float or settle depending on their viscosity.

e. Pumping nature of semi-solids or liquids food waste via the pipe using forced or natural flows.

f. Coating food waste increases the spreading of food waste as a single layer over another.

g. Cleaning of food waste from dryers or other processing units. For example, the removal of food waste soil from the equipment and pipeline surface.

h. Control of food waste processing parameters viz. velocity, the magnitude of the drop in pressure, design of piping, fluid transport system pumping requirement, power requirement for agitation, power requirement for blending and mixing, and amount of generated heat during extrusion, etc.

i. Influence on unit operations viz. mixing, grinding, sedimentation, filtration, evaporation, and drying separation, etc.

j. Select the best method of sorting and harvesting raw materials from food waste.

k. Select the best ingredients that can be used to process food waste.

l. Select the best technology/equipment needed to process food waste with desirable sensory and rheological properties.

m. Develop new forms of food waste (e.g. food waste from paneer, dietetic ice cream, low-fat mozzarella cheese, etc.)

n. Design packaging machines, processing equipment, and transportation system for food waste.

o. Determine flow properties and viscoelastic properties of food waste.

Case Studies: Role of Drying in Different Food Waste

Case Study A: Meat Waste

To evaluate the type of meat breed, its tenderness, to evaluate the effect of chilling, pickling, ageing, preservation, drying, etc. on rheological properties of meat waste for measurement of compactness and toughness of meat waste and meat waste product and establishment of quality for marketing and export, valorise high-protein animal waste (containing bones, meat, feather) for fertiliser purposes, the waste was processed by acid solubilisation and neutralised with potassium hydroxide solution, which yielded a liquid fertiliser with plant growth bio-stimulating properties (due to the presence of amino acids) (e.g. proteins).

Case Study B: Producing Natural Jams and Jelly Using Fruit or Vegetable Waste

Fruit and vegetable waste can be converted into natural jams and jelly using different drying methods to control the total solids. Referring to the interview regarding revival of food waste and turning it into jam, Wong and Feker won a category award in UT Food Lab's 2015 Food Challenge Prize converted banana waste into jam (*Interviewed: Revive Food on Turning Fruit Waste into Jam - Shareable*, n.d.).

Case Study C: Snacks Made from Food Waste Via Drying

Food waste from farmers can be transformed into snacks. Treasure8 is an organisation that collects food waste from farmers and transformed it into snacks, food enhancers, etc., in different forms (*Treasure8*, n.d.). Different drying methods will affect the product quality of the snack produced including the solids content and textural properties such as hardness, softness, and crispiness, and other properties. The other examples of upcycled snacks are CaPao cacao fruit bites, Renewal Mill oat chocolate chip cookie mix, and Pulp

Pantry salt "n" vinegar chips (*Are Upcycled Snacks Having Any Impact On Food Waste?*, n.d.)

Case Study D: Drying of Food Waste from Confectioneries

The confectionery products can be categorised as sugar confectionery (gelatine, sugar, and gum Arabic), milk chocolate confectionery (cocoa butter, cocoa liquor, and milk-based), dark chocolate confectionery (cocoa butter, cocoa liquor, and milk-based), milk chocolate biscuit confectionery (cocoa butter, cocoa liquor, and milk-based), and milk-based confectionery (milk-based, palm oil, and sugar) (Miah et al., 2018). Pilarska et al. (2018) produced biogas using confectionery waste, including chocolate bars (CB), wafers (W), and filled wafers (FW), by inoculation with digested cattle slurry and maize silage pulp. Suitable drying methods can be used to modify total solids.

Case Study E: Drying of Paste-Like Food Waste

Paste-like food waste is usually dried using a rotary dryer. The temperature of an exhaust gas from a drying chamber and moisture content is set in advance. Usually, an infrared moisture analyser is required to assist in determining the moisture content of the samples. Paste-like food waste is a semi-solid mixture such as spreads, tomato paste, relishes, gels, puddings, jams, and jellies, for example, tomato waste valorisation. It involves multiple steps and several heat treatments such as drying, hot break, and pasteurisation (Capanoglu et al., 2010).

Case Study F: Drying of Bakery Waste

Bakery waste is usually obtained from the recycling of the bakery industry or shops due to expire. It usually comes with packaging materials. It is required to be separated before processing or drying to convert into other materials. A rotary dryer is recommended for bakery waste. Bakery waste consists of clean and fermentable sugars and proteins. It can be converted into high-value chemicals, biofuels, bioplastics, and other renewable products with applications across many industries (Narisetty et al., 2021). Based on the availability, it can be valorised into fuel and enzymes. Table 2.2 summarised bread waste as feedstock to produce fuel and enzymes.

Case Study G: Drying of Dairy Product Waste

Diary product waste can be obtained from wastewater, expired feedstocks, expired products, contaminated products, kitchen waste, etc. The rich organic nature of dairy waste makes it a valuable feedstock to produce value-added products (Table 2.3). It has the potential to be developed into bioplastic, biomass, biohydrogen, bioethanol, and biosurfactant. For example, biomass can

TABLE 2.2

Microbial Production of Fuels, Chemicals, and Enzymes Using Bread Waste as a Feedstock

Feedstock	Microorganism	Product	Fermentation Mode	Titre	Yield[a]	Productivity	References
Waste bread hydrolysate	*Aspergillus* sp.	Glucoamylase	Submerged fermentation	-	8 units per g	-	Vu et al. (2018)
Waste bread hydrolysate	*Saccharomyces cerevisiae*	Ethanol	Separate hydrolysis and fermentation (SHF)	58 g/L	0.35 g/g	1.21 g/L.h	Dulf et al. (2016)
Waste bread hydrolysate	*Biohydrogenbacterium* R3	Hydrogen	Separate hydrolysis and fermentation (SHF)	7,482 mL	103 mL/g	103.91 mL/h	Wang et al. (2018)
Waste bread hydrolysate	*Saccharomyces cerevisiae*	Ethanol	Separate hydrolysis and fermentation (SHF)	33.9 g/L	0.25 g/g	0.36 g/L.h	Ballesteros-Vivas et al. (2019)
Waste bread hydrolysate	*Saccharomyces cerevisiae*	Ethanol	Separate hydrolysis and fermentation (SHF)	100 g/L	0.35 g/g	10 g/L.h	Fernández-Bolaños et al. (2006)
Waste bread	*Lactobacillus paracasei*	Lactic acid	Simultaneous saccharification and SSF (solid-state) fermentation)	28 g/L	0.056 g/g	0.58 g/L.h	Fernández-Bolaños et al. (2006)
Bread crumbs	*Thraustochytrium* sp. AH-2	Lipids	Submerged fermentation	390 mg/L	0.03 g/g	2.32 mg/L.h	Klinke et al. (2013)
Waste bread hydrolysate	*Bacillus amyloliquefaciens*	BDO+acetoin	Separate hydrolysis and fermentation (SHF)	103.9 g/L	0.39[b] g/g	0.87 g/L.h	Torres-León et al. (2016)
Waste bread	*Rhizopus oryzae*	a-Amylase	Solid-state fermentation (SSF)	-	100 units per g	-	Sánchez-Camargo et al. (2019)

(Continued)

TABLE 2.2 (Continued)

Microbial Production of Fuels, Chemicals, and Enzymes Using Bread Waste as a Feedstock

Feedstock	Microorganism	Product	Fermentation Mode	Titre	Yield[a]	Productivity	References
Waste bread hydrolysate	*Actinobacillus succinogenes*	Succinic acid	Separate hydrolysis and fermentation (SHF)	47.3 g/L	0.55 g/g	1.12 g/L.h	Ishisono et al. (2017)
Waste bread	*Rhizopus oryzae*	Protease	Solid-state fermentation (SSF)	-	2,400 units per g	-	Sánchez-Camargo et al. (2019)
Waste bread	Enzymatic hydrolysis (α-amylase+glucoamylase)+ biotransformation	Glucose–fructose syrup	Sequential hydrolysis and enzymatic biotransformation	-	0.45 g/g (glucose)+0.4 g/g	-	Martin-Rios et al. (2018)
Waste bread	*Aspergillus awamori* 2B.361 U2/1	Glucoamylase	Solid-state fermentation (SSF)	-	114 units per g	-	Morris et al. (2013)
Waste bread	*Aspergillus awamori* 2B.361 U2/1 (glucose isomerase)	Protease	Solid-state fermentation (SSF)	-	83.2 units per g (fructose)	-	Morris et al. (2013)
Waste bread hydrolysate	*Aspergillus* sp.	Protease	Submerged fermentation	-	117 units per g	-	Vu et al. (2018)

Source: Adapted from Narisetty et al. (2021).

[a] Yield: calculated per gram of waste bread saccharified for glucose production.

[b] Yield calculated per gram glucose.

TABLE 2.3

Bio-Based Products from Dairy Waste

Substrate	Product	Yield	Microorganism/ Biomolecule	References
Dairy wastewater	Bioplastic	7.23±0.08 g/L	*Rhodopseudomonas rutila*	Dinesh et al. (2020)
Dairy waste	Biomass	0.32 g/L	*Saccharomyces cereviesae*	Suman et al. (2017)
Dairy wastewater	Biohydrogen	113.2±2.9 mmol H_2/g COD	Landfill leachate sludge	Monari et al. (2019)
Scotta feedstock from ricotta cheese	Bioactive Peptide	60 g BSA eq/L and to 1.7 g AA eq/L	Papain and pancreatic enzymes	Wong et al. (2019)
Whey permeates	Bioethanol	14.9%–15.9% v/v	*Saccharomyces cerevisiae*	Parashar et al. (2016)
Dairy waste	Bioenergy (electricity)	850 mV and 28 μA	*Saccharomyces cereviesae*	Al-saned et al. (2021)
Panner (Cheese) whey	Biosurfactant	2.7–4.8 g/L	*Pseudomonas aeruginosa* SR17	Patowary et al. (2016)
Whey permeates	Bioethanol	0.32±0.007 g/g	*Saccharomyces cerevisiae*	Beniwal et al. (2021)
Dairy whey and vegetable oil waste	Biosurfactant	5.72 g/L	*Pseudomonas aeruginosa*	Ibrahim et al. (2021)

Source: Adapted from Usmani et al. (2022).

be further upgraded into char using different technologies. Drying technology can be applied during pre- or post-processes. The conversion of biomass into biofuel was a multi-step process. It plays an important role before the typical extraction process. The production of whey powder from cheese whey was using spray or drum drying.

2.5 Food Waste with Bioactive Compounds and Essential Oils

An overview of several important fruit-processing wastes, isolated bioactive compounds, and their potential functions, especially from popular exotic fruits. Figures 2.4 and 2.5 show that 107 bioactive compounds chemical structures can be isolated from food waste. In the previous text, some examples of waste generated by fruit processing at the food industry level are illustrated in Table 2.4.

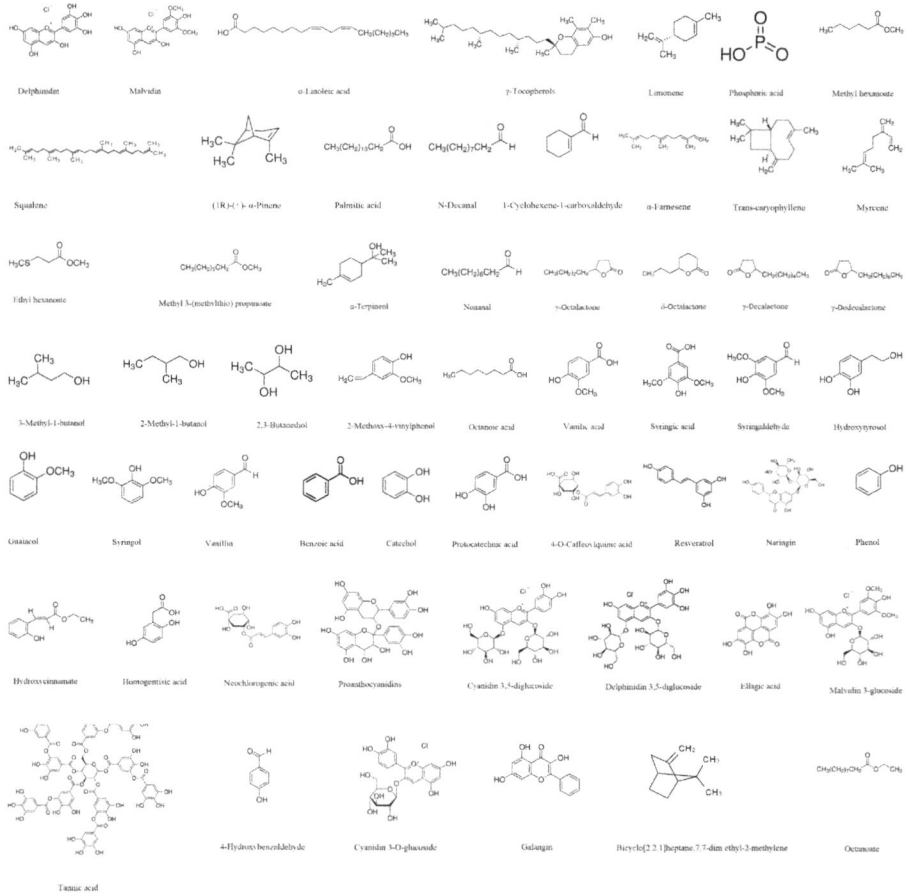

FIGURE 2.4
Bioactive compounds (no. 1–56) chemical structures which can be isolated from food waste.

2.6 Conclusions

Valorisation of food waste using drying technology is feasible for all kinds of characteristics exhibited by different food waste. It can also be used to stabilise or retain bioactive compounds of food waste with high nutritional values. Different compounds from food waste are presented in this chapter. The challenges of retaining valuable compounds, energy consumption, and selection of dryers will be discussed in Chapters 3 and 4.

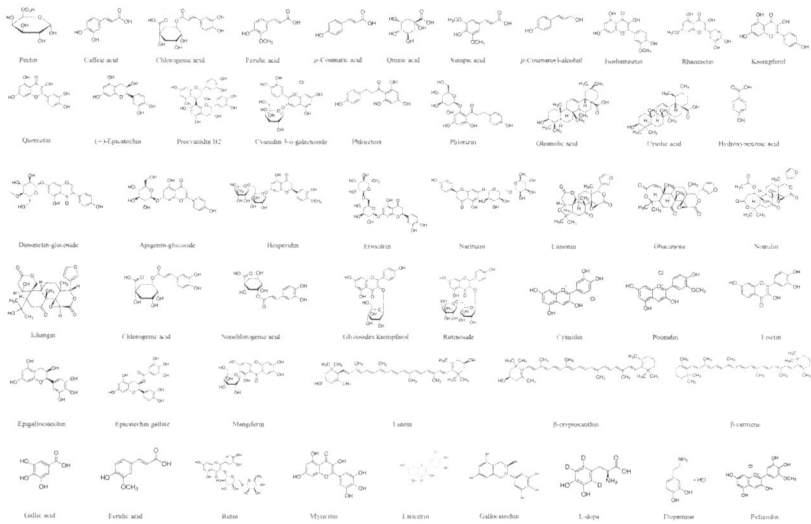

FIGURE 2.5
Bioactive compounds (no. 57–107) chemical structures which can be isolated from food waste.

TABLE 2.4

List of Food Waste with Bioactive Compounds and Essential Oils

Type of Food Waste	Bioactive Compound		Yield (mg/kg or %)	Bioactivity	References
	Class	Major Compounds			
Pomace of apple	Carbohydrates	Pectin, oligosaccharides	-	Dietary fibre, hypo-cholesterolemi, prebiotic	Ben-Othman et al. (2020), Lu and Foo (1997), Wang et al. (2020), Wolfe and Liu (2003), and Yeap Foo and Lu (1999)
	Phenolic acid	Caffeic acid, chlorogenic acid, ferulic acid, *p*-coumaric acid, quinic acid, sinapic acid *p*-coumaroyl-	523–1,542	Anti-oxidant, anti-inflammatory, anti-microbial, cardio-protective, anti-tumour	Barreira et al. (2019), Diñeiro García et al. (2009), and Lavelli and Corti (2011)
	Flavonoids	Isorhamnetin, rhamnetin, glycoconjugates, kaempferol, quercetin (−)-epicatechin, procyanidin B2	2,153–3,734		
	Anthyocanins	Cyanidin-3-*o*-galactoside	50–130		
	Dihydro-chalcones	Phloretein Phlorizin	688–2,535	Anti-diabetic. Anti-cholesterol. Promoting bone-growth blastogenesis.	Lavelli and Corti (2011), Shin et al. (2016), and Yuan et al. (2016)
	Triterpenoids	Oleanolic acid Ursolic acid	-	Anti-inflammatory, anti-microbial	Antika et al. (2017) and Yuan et al. (2016)

(*Continued*)

TABLE 2.4 (Continued)

List of Food Waste with Bioactive Compounds and Essential Oils

Type of Food Waste	Bioactive Compound		Yield (mg/kg or %)	Bioactivity	References
	Class	Major Compounds			
Peel of citrus fruit	Carbohydrates	Pectin	-	Dietary fibre, blood pressure lowering, reducing blood glucose prebiotic effect.	Anderson et al. (2009), Ishisono et al. (2017), Morris et al. (2013), and Naqash et al. (2017)
		Modified citrus pectin		Immuno-modulatory effect, anti-cancer	
Peel and pulp of citrus fruit	Phenolic acids	Caffeic acid Hydroxybenzoic acid	276–560	Anti-inflammatory, anti-oxidant, anti-cancer	Liu et al. (2019), Mahato et al. (2018), and Russo et al. (2014)
	Flavones	Diosmetin-glucoside Apigenin-glucoside	55–1,659		
	Flavanones	Hesperidin Eriocitrin Narirutin	10,646–22,298		
Seed of citrus fruit	Limonoids	Limonin, obacunone, nomilin Ichangin	375–114	Anti-cancer, anti-inflammatory, anti-bacterial activity, anti-oxidant	Sójka et al. (2015)
Pomace of plum	Phenolic acids	Chlorogenic acid Neochlorogenic acid	95.7	Anti-microbial, antioxidants, prevention of chronic diseases	Dulf et al. (2016)
	Flavonols	Glycosides kaempferol Quercetin Rutinoside	40.3		
Seed of plum	Anthocyanins	Glycosides Cyanidin Peonidin	6.5		González-García et al. (2014)
	Lipids	Essential oil rich in n–3 polyunsaturated fatty acids and sterol esters	53%		

(Continued)

TABLE 2.4 (*Continued*)

List of Food Waste with Bioactive Compounds and Essential Oils

Type of Food Waste	Bioactive Compound		Yield (mg/kg or %)	Bioactivity	References
	Class	Major Compounds			
Seeds of mango	Peptides	Bioactive peptides from protein hydrolysate	-	ACE inhibitory activity, anti-oxidant activity	Torres-León et al. (2016)
	Phenolic acids	Gallic acid and its derivatives	-	Anti-bacterial, anti-oxidant, anti-tumour, anti-viral, immune-modulatory effect.	Asif et al. (2016), Ballesteros-Vivas et al. (2019), and Wang et al. (2018)
	Flavonoids	Isoquercetin, quercetin Fisetin	7,200–13,000		
Mango peel	Catechins	Epigallocatechin Epicatechin Epicatechin gallate			
	Hydrolysable tannins		-		
	Xanthanoids	Mangiferin	13,600		Sánchez-Camargo et al. (2019) and Vu et al. (2018)
	Carotenoids		7.9		
	Carotenoids	Lutein β-cryptoxanthin, β-carotene	1,900	Prevention of age-related macular eye disease, anti-oxidant, regulation of bone homeostasis.	

(*Continued*)

TABLE 2.4 (*Continued*)

List of Food Waste with Bioactive Compounds and Essential Oils

Type of Food Waste	Bioactive Compound		Yield (mg/kg or %)	Bioactivity	References
	Class	Major Compounds			
Banana peel	Phenolic acids	P-coumaric acid Ferulic acid Sinapic acids Caffeic acid	99.5	Anti-bacterial, anti-fungal activity, anti-oxidant, reducing blood sugar, lowering cholesterol, the neuroprotective effect, anti-angiogenic activity.	Klavins et al. (2018)
	Flavonols	Rutin, myricitin quercetin kaempferol Laricitrin	1,019.6		
	Catechins	Catechin Gallocatechin Epicatechin	-		
	Catecholamines	L-dopa and dopamine	4,720		
Berries press residue of berries vaccinium genus berries (bilberries, blueberries, lingonberries, cranberries)	Anthocyanins	Cyanidin, petunidin Glycoconjugates of delphinidin, malvidin	27,890 (lingonberries) 84,120 (blueberry) 284,950 (bilberries) 43,530 (cranberry)	Prevention of various chronic diseases such as atherosclerosis, cardiovascular disease, and cancer.	Silva et al. (2017)
Branche waste of elderberry	Phenolic acids	Chlorogenic acid	45,600	Anti-inflammatory, anti-oxidant, anti-cancer	Yang et al. (2011)
	Flavonols	Quercetin and its glycoconjugates	468,200		
	Anthocyanins	Cyanidin and its glycoconjugates	2,530		

(*Continued*)

TABLE 2.4 (*Continued*)

List of Food Waste with Bioactive Compounds and Essential Oils

Type of Food Waste	Bioactive Compound		Yield (mg/kg or %)	Bioactivity	References
	Class	Major Compounds			
Seeds of wild and cultivated berries	Lipids	Essential oil rich in α-linoleic acid with a high content of α- and γ-tocopherols	14.61%–18.19%	Anti-oxidant, balancing diet fatty acid composition, regeneration of skin.	Mildner-Szkudlarz et al. (2019)
Essential oil from lemon peel waste		Limonene	13.21%		Hikal et al. (2022)
		Phosphoric acid, tribornyl ester	10.68%		
		Bicyclo[2.2.1]heptane,7,7-dim ethyl-2-methylene	6.61%		
		1,4-Methano-1H-indene,octahedron-1,7a-dimethyl-4-(1-methyl phenyl)	3.76%		
		Squalene			
		Bicyclo[2.2.1]heptan-2-ol,1,3,3-trimethyl-, (1S-endo)	3.19%		
		(1R)-α-Pinene	2.52%		
Essential oil from orange peel waste	Polyphenol	Limonene, narirutin, hesperidin, pectin	1.48%		

(*Continued*)

TABLE 2.4 (*Continued*)

List of Food Waste with Bioactive Compounds and Essential Oils

Type of Food Waste	Bioactive Compound Class	Major Compounds	Yield (mg/kg or %)	Bioactivity	References
Essential oil from pineapple peel and leaves waste	Esters, ketones, alcohols, aldehydes, acids	Tricin 4'O[10"O(8"hydroxyl) feruloyl acid; tricin; chrysoeriol; 1Op–(9'''Opcoumaroyl) glyceryl]		Inhibitory activities *Staphylococcus aureus* and *Escherichia coli*	Fernández-Bolaños et al. (2006)
		2,4-Coumaroylglycerol; c; 1-O-dichlorobenzoic coumaroylglycerol and 1, 3-O-diferuloylglycerol.			
		Palmitic acid	5.38%		
		n-decanal	0.95%		
		1cyclohexene-1-carboxaldehyde	4.27%		
		A-farnesene	1.26%		
		Trans-caryophyllene	0.53%		
		Myrcene	0.61%		
Essential oil from e pineapple pulp and core waste		Methyl hexanoate, ethyl hexanoate, methyl 3-(methylthio) propanoate, methyl octanoate, ethyl decanoate, α-terpineol, nonanal, and decanal,			
		γ-octalactone, δ-octalactone,	6.91%		
		γ-decalactone, and	4.75%		
		γ-dodecalactone,	8.35%		
		3-methyl-1-butanol	5.66%		
		2-methyl-1-butanol	4.29%		
		2,3-butanediol			
		2-methoxy-4-vinylphenol and Octanoic acid			

(*Continued*)

TABLE 2.4 (*Continued*)

List of Food Waste with Bioactive Compounds and Essential Oils

Type of Food Waste	Bioactive Compound		Yield (mg/kg or %)	Bioactivity	References
	Class	Major Compounds			
Olive whole stone	Phenolics	Vanilic acid, syringic acid, syringaldehyde, hydroxytyrosol			Klinke et al. (2013)
Olive seed husk		Vanillic acid, vanillin, syringic acid, hydroxytyrosol			
Wheat stows		Phenol, guaiacol, syringol, 4-hydroxybenzaldehyde, vanillin, syringic acid			Martin-Rios et al. (2018) and Takeoka et al. (2000)
Sugarcane bagasse		Benzoic acid, caffeic acid, catechol, p-coumaric acid, vanillic acid, protocatechuic acid			
Almond hulls		Chlorogenic acid, 4-O-caffeoylquinic acid			Takeoka et al. (2000)
Buckwheat hulls		Protocatechuic acid, hyperin, rutin			
Blueberry residue		Anthocyanins			Paes et al. (2014)
Apricot pomace		Carotenoids			Şanal et al. (2005)
Pistachio hull		Polyphenols			Goli et al. (2005)
Apricot by-products		β-Carotene			Şanal et al. (2005)
Guava seeds		Phenolic compounds			Castro-Vargas et al. (2010)
Grape peel		Anthocyanins, phenolic compounds			Ghafoor et al. (2011)
Grape by-products		Resveratrol			Casas et al. (2010)

(*Continued*)

TABLE 2.4 (*Continued*)

List of Food Waste with Bioactive Compounds and Essential Oils

Type of Food Waste	Bioactive Compound		Yield (mg/kg or %)	Bioactivity	References
	Class	Major Compounds			
Citrus peel		Naringin			Giannuzzo et al. (2003)
Red grape residue		Pro-anthocyanidins			Louli et al. (2004)
Tomato waste		Trans-lycopene			Nobre et al. (2009)
Spearmint leaves		Flavonoids			Bimakr et al. (2012)
Tea seed cake		Kaempferol glycosides			Li et al. (2010)
Green tea leaves		Catechins			Chang et al. (2000)
Carrot press cake		β-carotene			Vega et al. (1996)
Sweet potato waste		β-carotene and α-tocopherol			Okuno et al. (2002)
Tomato skin		Carotenoids			Shi et al. (2009)
Banana peel	Essential oil				Comim et al. (2010)
Peel of apple		Epicatechin, catechins, anthocyanins, quercitin Glycosides, chlorogenic acid, hydroxycinnamates			Wolfe and Liu (2003) and Yeap Foo and Lu (1999)
Avocado peel and seeds		Epicatechin, catechin, gallic acid, chlorogenic acid, cyanidin 3-glucoside, homogentisic acid			Deng et al. (2012)

(Continued)

TABLE 2.4 (*Continued*)

List of Food Waste with Bioactive Compounds and Essential Oils

Type of Food Waste	Bioactive Compound		Yield (mg/kg or %)	Bioactivity	References
	Class	Major Compounds			
Grape seeds		Coumaric acid, caffeic acid, ferulic acid, chlorogenic Acid, cinnamic acid, neochlorogenic acid, *P*-hydroxybenzoic acid, protocatechuic acid, vanillic acid, gallic acid, proanthocyanidins, quercetin 3-O-gluuronide, Quercetin, resveratrol			Maier et al. (2009), Negro et al. (2003) and Shrikhande (2000)
Guava skin and seeds		Catechin, cyanidin 3-glucoside, galangin, gallic acid, homogentisic acid, kaempferol			Deng et al. (2012)
Pomegranate peel and pericarp		Gallic acid, cyanidin-3,5-diglucoside, cyanidin-3-diglucoside, delphinidin-3,5-diglucoside			Gil et al. (2000) and Noda et al. (2002)
Mango kernel		Gallic acid, ellagic acid, gallates, gallotannins, Tannins			Arogba (2000) and Puravankara et al. (2000)
Pericarp and seeds of Litchi		Cyanidin-3-glucoside, cyanidin-3-rutonoside, malvidin-3-glucoside, gallic acid, epicatechin-3-gallate			Duan et al. (2007) and Lee and Wicker (1991)

Source: Adapted from Ben-Othman et al. (2020).

References

Agapios, A., Andreas, V., Marinos, S., Katerina, M., & Antonis, Z. A. (2020). Waste aroma profile in the framework of food waste management through household composting. *Journal of Cleaner Production, 257*, 120340. https://doi.org/10.1016/j.jclepro.2020.120340

Al-saned, A. J. O., Kitafa, B. A., & Badday, A. S. (2021). Microbial fuel cells (MFC) in the treatment of dairy wastewater. *IOP Conference Series: Materials Science and Engineering, 1067*(1), 012073. https://doi.org/10.1088/1757-899X/1067/1/012073

Anderson, J. W., Baird, P., Davis Jr, R. H., Ferreri, S., Knudtson, M., Koraym, A., Waters, V., & Williams, C. L. (2009). Health benefits of dietary fiber. *Nutrition Reviews, 67*(4), 188–205. https://doi.org/10.1111/j.1753-4887.2009.00189.x

Antika, L. D., Lee, E.-J., Kim, Y.-H., Kang, M.-K., Park, S.-H., Kim, D. Y., Oh, H., Choi, Y.-J., & Kang, Y.-H. (2017). Dietary phlorizin enhances osteoblastogenic bone formation through enhancing β-catenin activity via GSK-3β inhibition in a model of senile osteoporosis. *The Journal of Nutritional Biochemistry, 49*, 42–52. https://doi.org/10.1016/j.jnutbio.2017.07.014

Aqeela, N., Aji, S., Yaser, A. Z., Lamaming, J., Al, M., Ugak, M., Saalah, S., & Rajin, M. (2021). Production of food waste compost and its effect on the growth of Dwarf Crape Jasmine. *Jurnal Kejuruteraan, 33*(3), 413–424. https://doi.org/10.17576/jkukm-2021-33(3)-04

Are Upcycled Snacks Having any Impact on Food Waste? (n.d.). Retrieved January 27, 2023, from https://www.byrdie.com/upcycled-snacks-5196060

Arelli, V., Begum, S., Anupoju, G. R., Kuruti, K., & S., S. (2018). Dry anaerobic co-digestion of food waste and cattle manure: Impact of total solids, substrate ratio and thermal pre treatment on methane yield and quality of biomanure. *Bioresource Technology, 253*, 273–280. https://doi.org/10.1016/j.biortech.2018.01.050

Arogba, S. S. (2000). Mango (Mangifera indica) Kernel: Chromatographic analysis of the tannin, and stability study of the associated polyphenol oxidase activity. *Journal of Food Composition and Analysis, 13*(2), 149–156. https://doi.org/10.1006/jfca.1999.0838

Asif, A., Farooq, U., Akram, K., Hayat, Z., Shafi, A., Sarfraz, F., Sidhu, M. A. I., Rehman, H., & Aftab, S. (2016). Therapeutic potentials of bioactive compounds from mango fruit wastes. *Trends in Food Science & Technology, 53*, 102–112. https://doi.org/10.1016/j.tifs.2016.05.004

Ballesteros-Vivas, D., Álvarez-Rivera, G., Morantes, S. J., del P. Sánchez-Camargo, A., Ibáñez, E., Parada-Alfonso, F., & Cifuentes, A. (2019). An integrated approach for the valorization of mango seed kernel: Efficient extraction solvent selection, phytochemical profiling and antiproliferative activity assessment. *Food Research International, 126*, 108616. https://doi.org/10.1016/j.foodres.2019.108616

Barreira, J. C. M., Arraibi, A. A., & Ferreira, I. C. F. R. (2019). Bioactive and functional compounds in apple pomace from juice and cider manufacturing: Potential use in dermal formulations. *Trends in Food Science & Technology, 90*, 76–87. https://doi.org/10.1016/j.tifs.2019.05.014

Beniwal, A., Saini, P., De, S., & Vij, S. (2021). Harnessing the nutritional potential of concentrated whey for enhanced galactose flux in fermentative yeast. *LWT, 141*, 110840. https://doi.org/10.1016/j.lwt.2020.110840

Ben-Othman, S., Jõudu, I., & Bhat, R. (2020). Bioactives from agri-food wastes: Present insights and future challenges. *Molecules*, *25*(3), 510. https://doi.org/10.3390/molecules25030510

Bimakr, M., Rahman, R. A., Ganjloo, A., Taip, F. S., Salleh, L. M., & Sarker, M. Z. I. (2012). Optimization of supercritical carbon dioxide extraction of bioactive flavonoid compounds from spearmint (Mentha spicata L.) leaves by using response surface methodology. *Food and Bioprocess Technology*, *5*(3), 912–920. https://doi.org/10.1007/s11947-010-0504-4

Capanoglu, E., Beekwilder, J., Boyacioglu, D., de Vos, R. C. H., & Hall, R. D. (2010). The effect of industrial food processing on potentially health-beneficial tomato antioxidants. *Critical Reviews in Food Science and Nutrition*, *50*(10), 919–930. https://doi.org/10.1080/10408390903001503

Casas, L., Mantell, C., Rodríguez, M., Ossa, E. J. M. de la, Roldán, A., Ory, I. de, Caro, I., & Blandino, A. (2010). Extraction of resveratrol from the pomace of Palomino fino grapes by supercritical carbon dioxide. *Journal of Food Engineering*, *96*(2), 304–308. https://doi.org/10.1016/j.jfoodeng.2009.08.002

Castro-Vargas, H. I., Rodríguez-Varela, L. I., Ferreira, S. R. S., & Parada-Alfonso, F. (2010). Extraction of phenolic fraction from guava seeds (Psidium guajava L.) using supercritical carbon dioxide and co-solvents. *The Journal of Supercritical Fluids*, *51*(3), 319–324. https://doi.org/10.1016/j.supflu.2009.10.012

Chang, C. J., Chiu, K.-L., Chen, Y.-L., & Chang, C.-Y. (2000). Separation of catechins from green tea using carbon dioxide extraction. *Food Chemistry*, *68*(1), 109–113. https://doi.org/10.1016/S0308-8146(99)00176-4

Comim, S. R. R., Madella, K., Oliveira, J. V., & Ferreira, S. R. S. (2010). Supercritical fluid extraction from dried banana peel (Musa spp., genomic group AAB): Extraction yield, mathematical modeling, economical analysis and phase equilibria. *The Journal of Supercritical Fluids*, *54*(1), 30–37. https://doi.org/10.1016/j.supflu.2010.03.010

Deng, G.-F., Shen, C., Xu, X.-R., Kuang, R.-D., Guo, Y.-J., Zeng, L.-S., Gao, L.-L., Lin, X., Xie, J.-F., Xia, E.-Q., Li, S., Wu, S., Chen, F., Ling, W.-H., & Li, H.-B. (2012). Potential of fruit wastes as natural resources of bioactive compounds. *International Journal of Molecular Sciences*, *13*(7), 8308–8323. https://doi.org/10.3390/ijms13078308

Diñeiro García, Y., Valles, B. S., & Picinelli Lobo, A. (2009). Phenolic and antioxidant composition of by-products from the cider industry: Apple pomace. *Food Chemistry*, *117*(4), 731–738. https://doi.org/10.1016/j.foodchem.2009.04.049

Dinesh, G. H., Nguyen, D. D., Ravindran, B., Chang, S. W., Vo, D.-V. N., Bach, Q.-V., Tran, H. N., Basu, M. J., Mohanrasu, K., Murugan, R. S., Swetha, T. A., Sivapraksh, G., Selvaraj, A., & Arun, A. (2020). Simultaneous biohydrogen (H₂) and bioplastic (poly-β-hydroxybutyrate-PHB) productions under dark, photo, and subsequent dark and photo fermentation utilizing various wastes. *International Journal of Hydrogen Energy*, *45*(10), 5840–5853. https://doi.org/10.1016/j.ijhydene.2019.09.036

Donahue, D. W., Chalmers, J. A., & Storey, J. A. (1998). Evaluation of in-vessel composting of university postconsumer food wastes. *Compost Science & Utilization*, *6*(2), 75–81. https://doi.org/10.1080/1065657X.1998.10701922

Duan, X., Jiang, Y., Su, X., Zhang, Z., & Shi, J. (2007). Antioxidant properties of anthocyanins extracted from litchi (Litchi chinenesis Sonn.) fruit pericarp tissues in relation to their role in the pericarp browning. *Food Chemistry*, *101*(4), 1365–1371. https://doi.org/10.1016/j.foodchem.2005.06.057

Dulf, F. V., Vodnar, D. C., & Socaciu, C. (2016). Effects of solid-state fermentation with two filamentous fungi on the total phenolic contents, flavonoids, antioxidant activities and lipid fractions of plum fruit (Prunus domestica L.) by-products. *Food Chemistry, 209*, 27–36. https://doi.org/10.1016/j.foodchem.2016.04.016

Fernández-Bolaños, J., Rodríguez, G., Rodríguez, R., Guillén, R., & Jiménez, A. (2006). Extraction of interesting organic compounds from olive oil waste. *Grasas y Aceites, 57*(1). https://doi.org/10.3989/gya.2006.v57.i1.25

Ghafoor, K., Hui, T., & Choi, Y. H. (2011). Optimization of ultrasonic-assisted extraction of total anthocyanins from grape peel using response surface methodology. *Journal of Food Biochemistry, 35*(3), 735–746. https://doi.org/10.1111/j.1745-4514.2010.00413.x

Giannuzzo, A. N., Boggetti, H. J., Nazareno, M. A., & Mishima, H. T. (2003). Supercritical fluid extraction of naringin from the peel of Citrus paradisi. *Phytochemical Analysis, 14*(4), 221–223. https://doi.org/10.1002/pca.706

Gil, M. I., Tomás-Barberán, F. A., Hess-Pierce, B., Holcroft, D. M., & Kader, A. A. (2000). Antioxidant activity of pomegranate juice and its relationship with phenolic composition and processing. *Journal of Agricultural and Food Chemistry, 48*(10), 4581–4589. https://doi.org/10.1021/jf000404a

Goli, A. H., Barzegar, M., & Sahari, M. A. (2005). Antioxidant activity and total phenolic compounds of pistachio (Pistachia vera) hull extracts. *Food Chemistry, 92*(3), 521–525. https://doi.org/10.1016/j.foodchem.2004.08.020

González-García, E., Marina, M. L., & García, M. C. (2014). Plum (Prunus Domestica L.) by-product as a new and cheap source of bioactive peptides: Extraction method and peptides characterization. *Journal of Functional Foods, 11*, 428–437. https://doi.org/10.1016/j.jff.2014.10.020

Hikal, W. M., Said-Al Ahl, H. A. H., Tkachenko, K. G., Bratovcic, A., Szczepanek, M., & Rodriguez, R. M. (2022). Sustainable and environmentally friendly essential oils extracted from pineapple waste. *Biointerface Research in Applied Chemistry, 12*(5), 6833–6844. https://doi.org/10.33263/BRIAC125.68336844

Ibrahim, M. A., El-Araby, R., Abdelkader, E., Abdelsalam,; A M, & Ismail, E. H. (2021). Fuel range hydrocarbon synthesized by hydrocracking of waste cooking oil via Co/Zn-Al$_2$O$_4$ nano particles. *Egyptian Journal of Applied Science, 36*(9), 43–55. https://doi.org/10.21608/ejas.2021.220043

Ishisono, K., Yabe, T., & Kitaguchi, K. (2017). Citrus pectin attenuates endotoxin shock via suppression of Toll-like receptor signaling in Peyer's patch myeloid cells. *The Journal of Nutritional Biochemistry, 50*, 38–45. https://doi.org/10.1016/j.jnutbio.2017.07.016

Jiang, Y., Dennehy, C., Lawlor, P. G., Hu, Z., McCabe, M., Cormican, P., Zhan, X., & Gardiner, G. E. (2018). Inhibition of volatile fatty acids on methane production kinetics during dry co-digestion of food waste and pig manure. *Waste Management, 79*, 302–311. https://doi.org/10.1016/j.wasman.2018.07.049

Kamaruddin, M. A., Jantira, N. N., & Alrozi, R. (2020). Food waste quantification and characterization as a measure towards effective food waste management in university. *IOP Conference Series: Materials Science and Engineering, 743*(1). https://doi.org/10.1088/1757-899X/743/1/012041

Khalida, A., Arumugam, V., Abdullah, L. C., Manaf, L. A., & Ismail, M. H. (2022). Dehydrated food waste for composting: An overview. *Pertanika Journal of Science and Technology, 30*(4), 2933–2960. https://doi.org/10.47836/pjst.30.4.33

Kim, M.-G., Sivagurunathan, P., Lee, M.-K., Im, S., Shin, S.-R., Choi, C.-S., & Kim, D.-H. (2019). Rheological properties of hydrogen fermented food waste. *International Journal of Hydrogen Energy, 44*(4), 2239–2245. https://doi.org/10.1016/j.ijhydene.2018.07.073

Klavins, L., Kviesis, J., Nakurte, I., & Klavins, M. (2018). Berry press residues as a valuable source of polyphenolics: Extraction optimisation and analysis. *LWT, 93*, 583–591. https://doi.org/10.1016/j.lwt.2018.04.021

Klinke, M. E., Wilson, M. E., Hafsteinsdóttir, T. B., & Jónsdóttir, H. (2013). Recognizing new perspectives in eating difficulties following stroke: A concept analysis. *Disability and Rehabilitation, 35*(17), 1491–1500. https://doi.org/10.3109/09638288.2012.736012

Lavelli, V., & Corti, S. (2011). Phloridzin and other phytochemicals in apple pomace: Stability evaluation upon dehydration and storage of dried product. *Food Chemistry, 129*(4), 1578–1583. https://doi.org/10.1016/j.foodchem.2011.06.011

Lee, H. S., & Wicker, L. (1991). Anthocyanin pigments in the skin of lychee fruit. *Journal of Food Science, 56*(2), 466–468. https://doi.org/10.1111/j.1365-2621.1991.tb05305.x

Li, B., Xu, Y., Jin, Y.-X., Wu, Y.-Y., & Tu, Y.-Y. (2010). Response surface optimization of supercritical fluid extraction of kaempferol glycosides from tea seed cake. *Industrial Crops and Products, 32*(2), 123–128. https://doi.org/10.1016/j.indcrop.2010.04.002

Lin, C. S. K., Pfaltzgraff, L. A., Herrero-Davila, L., Mubofu, E. B., Abderrahim, S., Clark, J. H., Koutinas, A. A., Kopsahelis, N., Stamatelatou, K., Dickson, F., Thankappan, S., Mohamed, Z., Brocklesby, R., & Luque, R. (2013). Food waste as a valuable resource for the production of chemicals, materials and fuels. Current situation and global perspective. *Energy & Environmental Science, 6*(2), 426. https://doi.org/10.1039/c2ee23440h

Liu, S., Zhang, S., Lv, X., Lu, J., Ren, C., Zeng, Z., Zheng, L., Zhou, X., Fu, H., Zhou, D., & Chen, Y. (2019). Limonin ameliorates ulcerative colitis by regulating STAT3/miR-214 signaling pathway. *International Immunopharmacology, 75*, 105768. https://doi.org/10.1016/j.intimp.2019.105768

Louli, V., Ragoussis, N., & Magoulas, K. (2004). Recovery of phenolic antioxidants from wine industry by-products. *Bioresource Technology, 92*(2), 201–208. https://doi.org/10.1016/j.biortech.2003.06.002

Lu, Y., & Foo, L. Y. (1997). Identification and quantification of major polyphenols in apple pomace. *Food Chemistry, 59*(2), 187–194. https://doi.org/10.1016/S0308-8146(96)00287-7

Mahato, N., Sharma, K., Sinha, M., & Cho, M. H. (2018). Citrus waste derived nutra-/pharmaceuticals for health benefits: Current trends and future perspectives. *Journal of Functional Foods, 40*, 307–316. https://doi.org/10.1016/j.jff.2017.11.015

Maier, T., Schieber, A., Kammerer, D. R., & Carle, R. (2009). Residues of grape (Vitis vinifera L.) seed oil production as a valuable source of phenolic antioxidants. *Food Chemistry, 112*(3), 551–559. https://doi.org/10.1016/j.foodchem.2008.06.005

Martin-Rios, C., Demen-Meier, C., Gössling, S., & Cornuz, C. (2018). Food waste management innovations in the foodservice industry. *Waste Management, 79*, 196–206. https://doi.org/10.1016/j.wasman.2018.07.033

Miah, J. H., Griffiths, A., McNeill, R., Halvorson, S., Schenker, U., Espinoza-Orias, N. D., Morse, S., Yang, A., & Sadhukhan, J. (2018). Environmental management of confectionery products: Life cycle impacts and improvement strategies. *Journal of Cleaner Production*, *177*, 732–751. https://doi.org/10.1016/j.jclepro.2017.12.073

Mildner-Szkudlarz, S., Różańska, M., Siger, A., Kowalczewski, P. Ł., & Rudzińska, M. (2019). Changes in chemical composition and oxidative stability of cold-pressed oils obtained from by-product roasted berry seeds. *LWT*, *111*, 541–547. https://doi.org/10.1016/j.lwt.2019.05.080

Monari, S., Ferri, M., Russo, C., Prandi, B., Tedeschi, T., Bellucci, P., Zambrini, A. V., Donati, E., & Tassoni, A. (2019). Enzymatic production of bioactive peptides from scotta, an exhausted by-product of ricotta cheese processing. *PLOS One*, *14*(12), e0226834. https://doi.org/10.1371/journal.pone.0226834

Morris, V. J., Belshaw, N. J., Waldron, K. W., & Maxwell, E. G. (2013). The bioactivity of modified pectin fragments. *Bioactive Carbohydrates and Dietary Fibre*, *1*(1), 21–37. https://doi.org/10.1016/j.bcdf.2013.02.001

Naqash, F., Masoodi, F. A., Rather, S. A., Wani, S. M., & Gani, A. (2017). Emerging concepts in the nutraceutical and functional properties of pectin—A review. *Carbohydrate Polymers*, *168*, 227–239. https://doi.org/10.1016/j.carbpol.2017.03.058

Narisetty, V., Cox, R., Willoughby, N., Aktas, E., Tiwari, B., Matharu, A. S., Salonitis, K., & Kumar, V. (2021). Recycling bread waste into chemical building blocks using a circular biorefining approach. *Sustainable Energy & Fuels*, *5*(19), 4842–4849. https://doi.org/10.1039/D1SE00575H

Negro, C., Tommasi, L., & Miceli, A. (2003). Phenolic compounds and antioxidant activity from red grape marc extracts. *Bioresource Technology*, *87*(1), 41–44. https://doi.org/10.1016/S0960-8524(02)00202-X

Nobre, B. P., Palavra, A. F., Pessoa, F. L. P., & Mendes, R. L. (2009). Supercritical CO_2 extraction of trans-lycopene from Portuguese tomato industrial waste. *Food Chemistry*, *116*(3), 680–685. https://doi.org/10.1016/j.foodchem.2009.03.011

Noda, Y., Kaneyuki, T., Mori, A., & Packer, L. (2002). Antioxidant activities of pomegranate fruit extract and its anthocyanidins: Delphinidin, cyanidin, and pelargonidin. *Journal of Agricultural and Food Chemistry*, *50*(1), 166–171. https://doi.org/10.1021/jf0108765

Okuno, S., Yoshinaga, M., Nakatani, M., Ishiguro, K., Yoshimoto, M., Morishita, T., Uehara, T., & Kawano, M. (2002). Extraction of antioxidants in sweetpotato waste powder with supercritical carbon dioxide. *Food Science and Technology Research*, *8*(2), 154–157. https://doi.org/10.3136/fstr.8.154

Paes, J., Dotta, R., Barbero, G. F., & Martínez, J. (2014). Extraction of phenolic compounds and anthocyanins from blueberry (Vaccinium myrtillus L.) residues using supercritical CO_2 and pressurized liquids. *The Journal of Supercritical Fluids*, *95*, 8–16. https://doi.org/10.1016/j.supflu.2014.07.025

Parashar, A., Jin, Y., Mason, B., Chae, M., & Bressler, D. C. (2016). Incorporation of whey permeate, a dairy effluent, in ethanol fermentation to provide a zero waste solution for the dairy industry. *Journal of Dairy Science*, *99*(3), 1859–1867. https://doi.org/10.3168/jds.2015-10059

Patowary, R., Patowary, K., Kalita, M. C., & Deka, S. (2016). Utilization of paneer whey waste for cost-effective production of rhamnolipid biosurfactant. *Applied Biochemistry and Biotechnology*, *180*(3), 383–399. https://doi.org/10.1007/s12010-016-2105-9

Pilarska, A. A., Pilarski, K., Wolna-Maruwka, A., Boniecki, P., & Zaborowicz, M. (2018). Use of confectionery waste in biogas production by the anaerobic digestion process. *Molecules*, *24*(1), 37. https://doi.org/10.3390/molecules24010037

Puravankara, D., Boghra, V., & Sharma, R. (2000). Effect of antioxidant principles isolated from mango (Mangifera indica L) seed kernels on oxidative stability of buffalo ghee (butter-fat). *Journal of the Science of Food and Agriculture, 80*(4), 522–526. https://doi.org/10.1002/(SICI)1097-0010(200003)80:4<522::AID-JSFA560>3.0.CO;2-R

Razali, Z. B., Rahim, A., & Hasim, A. (2013). Conceptual Design and Analysis of Multi-angle Kitchen Waste Grater Mechanism for Biodegrading Kitchen Waste Green Environmental Product Design View project Sustainable Practical Intelligence in Higher Engineering Education View project Conceptual Design and Analysis of Multi-angle Kitchen Waste Grater Mechanism for Biodegrading Kitchen Waste. *Final Year Project.* https://doi.org/10.13140/RG.2.1.3832.6249

Russo, M., Bonaccorsi, I., Torre, G., Sarò, M., Dugo, P., & Mondello, L. (2014). Underestimated sources of flavonoids, limonoids and dietary fibre: Availability in lemon's by-products. *Journal of Functional Foods, 9*, 18–26. https://doi.org/10.1016/j.jff.2014.04.004

Şanal, İ. S., Bayraktar, E., Mehmetoğlu, Ü., & Çalımlı, A. (2005). Determination of optimum conditions for SC-(CO$_2$+ethanol) extraction of β-carotene from apricot pomace using response surface methodology. *The Journal of Supercritical Fluids, 34*(3), 331–338. https://doi.org/10.1016/j.supflu.2004.08.005

Sánchez-Camargo, A. del P., Gutiérrez, L.-F., Vargas, S. M., Martinez-Correa, H. A., Parada-Alfonso, F., & Narváez-Cuenca, C.-E. (2019). Valorisation of mango peel: Proximate composition, supercritical fluid extraction of carotenoids, and application as an antioxidant additive for an edible oil. *The Journal of Supercritical Fluids, 152*, 104574. https://doi.org/10.1016/j.supflu.2019.104574

Selvam, A., Ilamathi, P. M. K., Udayakumar, M., Murugesan, K., Banu, J. R., Khanna, Y., & Wong, J. (2021). Food Waste Properties. In Wong, J. , Kaur, G., Taherzadeh, M, Pandey, A., & Lasaridi, K. (Eds.), *Current Developments in Biotechnology and Bioengineering: Sustainable Food Waste Management: Resource Recovery and Treatment* (pp. 11–41). Elsevier. https://doi.org/10.1016/B978-0-12-819148-4.00002-6

Shareable. *Interviewed: Revive Food on Turning Fruit Waste into Jam.* (n.d.). Retrieved January 30, 2023, from https://www.shareable.net/interviewed-revive-food-on-turning-fruit-waste-into-jam/

Shi, J., Khatri, M., Xue, S. J., Mittal, G. S., Ma, Y., & Li, D. (2009). Solubility of lycopene in supercritical CO$_2$ fluid as affected by temperature and pressure. *Separation and Purification Technology, 66*(2), 322–328. https://doi.org/10.1016/j.seppur.2008.12.012

Shi, S., Zou, D., Wang, Q., Xia, X., Zheng, T., Wu, C., & Gao, M. (2016). Responses of ammonia-oxidizing bacteria community composition to temporal changes in physicochemical parameters during food waste composting. *RSC Advances, 6*(12), 9541–9548. https://doi.org/10.1039/C5RA22067J

Shin, S.-K., Cho, S.-J., Jung, U., Ryu, R., & Choi, M.-S. (2016). Phlorizin supplementation attenuates obesity, inflammation, and hyperglycemia in diet-induced obese mice fed a high-fat diet. *Nutrients, 8*(2), 92. https://doi.org/10.3390/nu8020092

Shrikhande, A. J. (2000). Wine by-products with health benefits. *Food Research International, 33*(6), 469–474. https://doi.org/10.1016/S0963-9969(00)00071-5

Silva, P., Ferreira, S., & Nunes, F. M. (2017). Elderberry (Sambucus nigra L.) by-products a source of anthocyanins and antioxidant polyphenols. *Industrial Crops and Products, 95*, 227–234. https://doi.org/10.1016/j.indcrop.2016.10.018

Sójka, M., Kołodziejczyk, K., Milala, J., Abadias, M., Viñas, I., Guyot, S., & Baron, A. (2015). Composition and properties of the polyphenolic extracts obtained from industrial plum pomaces. *Journal of Functional Foods, 12,* 168–178. https://doi.org/10.1016/j.jff.2014.11.015

Suman, G., Nupur, M., Anuradha, S., & Pradeep, B. (2017). Characterization of dairy waste and its utilisation as substrate for production of single cell protein. *IOSR Journal of Biotechnology and Biochemistry (IOSR-JBB, 3*(4), 73–78. https://doi.org/10.9790/264X-03047378

Takeoka, G., Dao, L., Teranishi, R., Wong, R., Flessa, S., Harden, L., & Edwards, R. (2000). Identification of three triterpenoids in almond hulls. *Journal of Agricultural and Food Chemistry, 48*(8), 3437–3439. https://doi.org/10.1021/jf9908289

Torres-León, C., Rojas, R., Contreras-Esquivel, J. C., Serna-Cock, L., Belmares-Cerda, R. E., & Aguilar, C. N. (2016). Mango seed: Functional and nutritional properties. *Trends in Food Science & Technology, 55,* 109–117. https://doi.org/10.1016/j.tifs.2016.06.009

Treasure8. (n.d.). Retrieved January 30, 2023, from https://www.treasure8.com/

Usmani, Z., Sharma, M., Gaffey, J., Sharma, M., Dewhurst, R. J., Moreau, B., Newbold, J., Clark, W., Thakur, V. K., & Gupta, V. K. (2022). Valorization of dairy waste and by-products through microbial bioprocesses. *Bioresource Technology, 346,* 126444. https://doi.org/10.1016/j.biortech.2021.126444

Vega, P. J., Balaban, M. O., Sims, C. A., O'Keefe, S. F., & Cornell, J. A. (1996). Supercritical carbon dioxide extraction efficiency for carotenes from carrots by RSM. *Journal of Food Science, 61*(4), 757–759. https://doi.org/10.1111/j.1365-2621.1996.tb12198.x

Verma, M. van den B., de Vreede, L., Achterbosch, T., & Rutten, M. M. (2020). Consumers discard a lot more food than widely believed: Estimates of global food waste using an energy gap approach and affluence elasticity of food waste. *PLOS One, 15*(2), e0228369. https://doi.org/10.1371/journal.pone.0228369

Vu, H. T., Scarlett, C. J., & Vuong, Q. V. (2018). Phenolic compounds within banana peel and their potential uses: A review. *Journal of Functional Foods, 40,* 238–248. https://doi.org/10.1016/j.jff.2017.11.006

Wang, X., Gao, L., Lin, H., Song, J., Wang, J., Yin, Y., Zhao, J., Xu, X., Li, Z., & Li, L. (2018). Mangiferin prevents diabetic nephropathy progression and protects podocyte function via autophagy in diabetic rat glomeruli. *European Journal of Pharmacology, 824,* 170–178. https://doi.org/10.1016/j.ejphar.2018.02.009

Wang, Z., Xu, B., Luo, H., Meng, K., Wang, Y., Liu, M., Bai, Y., Yao, B., & Tu, T. (2020). Production pectin oligosaccharides using Humicola insolens Y1-derived unusual pectate lyase. *Journal of Bioscience and Bioengineering, 129*(1), 16–22. https://doi.org/10.1016/j.jbiosc.2019.07.005

Wolfe, K. L., & Liu, R. H. (2003). Apple peels as a value-added food ingredient. *Journal of Agricultural and Food Chemistry, 51*(6), 1676–1683. https://doi.org/10.1021/jf025916z

Wong, Y. M., Show, P. L., Wu, T. Y., Leong, H. Y., Ibrahim, S., & Juan, J. C. (2019). Production of bio-hydrogen from dairy wastewater using pretreated landfill leachate sludge as an inoculum. *Journal of Bioscience and Bioengineering, 127*(2), 150–159. https://doi.org/10.1016/j.jbiosc.2018.07.012

Wu, L.-J., Kobayashi, T., Kuramochi, H., Li, Y.-Y., & Xu, K.-Q. (2016). Improved biogas production from food waste by co-digestion with de-oiled grease trap waste. *Bioresource Technology, 201,* 237–244. https://doi.org/10.1016/j.biortech.2015.11.061

Yang, B., Ahotupa, M., Määttä, P., & Kallio, H. (2011). Composition and antioxidative activities of supercritical CO_2-extracted oils from seeds and soft parts of northern berries. *Food Research International*, 44(7), 2009–2017. https://doi.org/10.1016/j.foodres.2011.02.025

Yeap Foo, L., & Lu, Y. (1999). Isolation and identification of procyanidins in apple pomace. *Food Chemistry*, 64(4), 511–518. https://doi.org/10.1016/S0308-8146(98)00150-2

Yuan, X., Wu, Y., Liang, J., Yuan, H., Zhao, X., Zhu, D., Liu, H., Lin, J., Huang, S., Lai, X., Chen, S., & Hou, S. (2016). Phlorizin treatment attenuates obesity and related disorders through improving BAT thermogenesis. *Journal of Functional Foods*, 27, 429–438. https://doi.org/10.1016/j.jff.2016.09.022

3

Drying Characteristics, Kinetics, and Different Drying Techniques of Different Food Wastes

3.1 Drying Characteristics

Drying of food waste is defined as the vapourization and removal of water from solids, and mixture to form dry matter that involved simultaneous heat and mass transfer. Drying occurs when heat is supplied to wet food waste. During drying, heat transfer occurred, and the heat transferred from the other outer parts of the food waste to the internal part of the food waste via convection, conduction, radiation, and radio frequency. A typical drying characteristic of food waste is shown in Figure 3.1. The drying process started with the initial transient period followed by a constant rate period. When the heat was transferred to the internal parts of the food waste samples, it will exhibit the first falling rate period followed by the second falling rate period. The only way to change the drying characteristics is by changing the heat source after the first falling rate period.

In drying food waste, it is important to include TGA and DTG plots. This TGA plot shows the decomposition of food waste in air at different heating rates; meanwhile the DTG provides the decomposition rate for evaluating the mass-loss steps. The decomposition occurs in three mass-loss steps: the combustion of volatiles, the transition stage, combustion, and burnout of fixed carbon. The TGA plot can provide information on the total percentage of mass loss at different stages from water loss to carbon loss. Examples of typical combustion characteristics of vegetable leaves and steamed rice are shown in Figures 3.2 and 3.3. According to Liu et al. (2016), the vegetable leaves showed two significant weight loss peaks over two temperature sections and steamed rice had two weight loss peaks as vegetable leaves have high volatile content, but steamed rice has low fixed carbon content. These two samples are easier to ignite compared to sub-bituminous coal. Microwave drying reduces the ignition temperature of food waste where the higher the intensity the lower the ignition temperature due to the evaporation of water

DOI: 10.1201/9781003312802-3

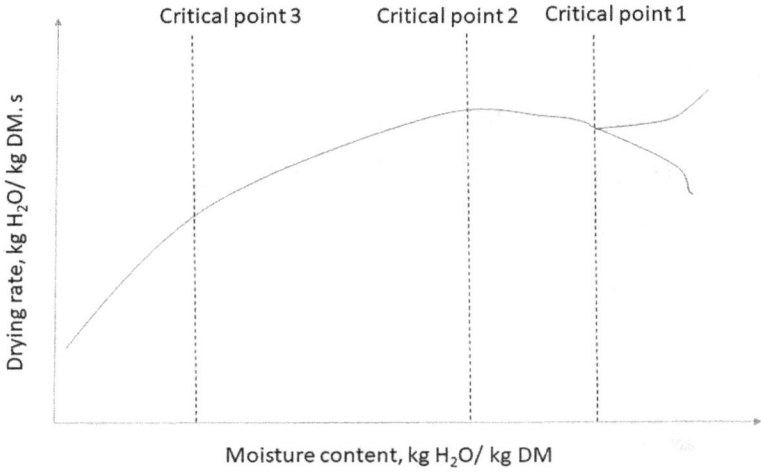

FIGURE 3.1
Typical drying characteristics and kinetics curve for food waste.

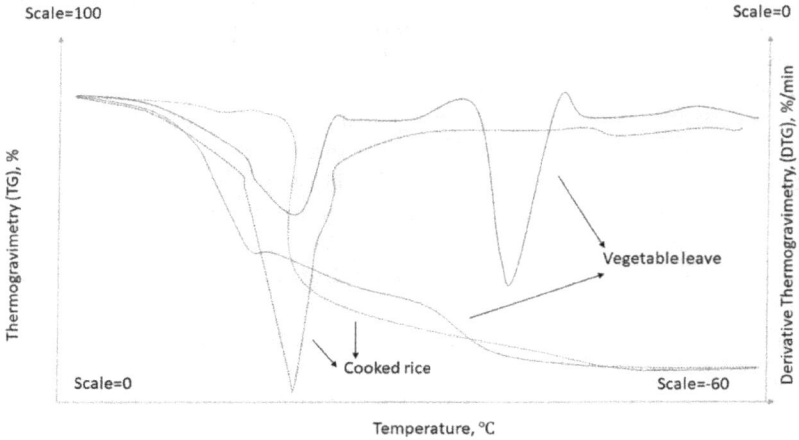

FIGURE 3.2
TGA and DTG plots for electrical drying and microwave drying of food waste. Adapted from Liu et al. (2016).

from the waste and expanded internal pores but not significant in cooked rice. In terms of combustion rate, it increases with the microwave intensity.

3.2 Effective Diffusivities of Food Waste

Different drying technologies can change the effective diffusivities of water from food waste samples. It has a higher impact than the operating

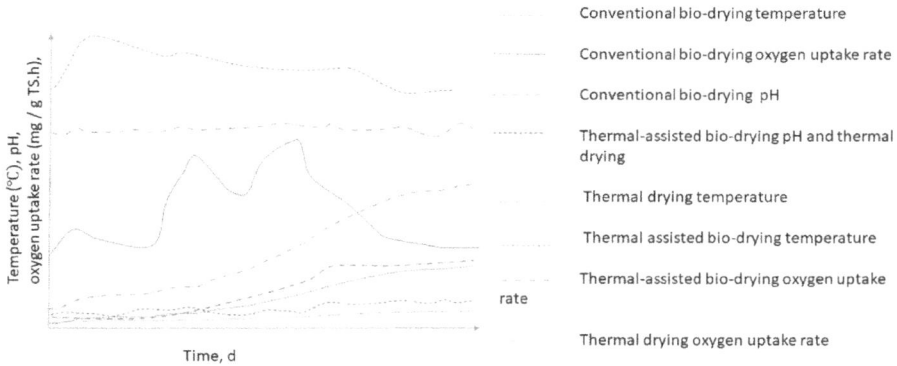

Conventional bio-drying temperature

Conventional bio-drying oxygen uptake rate

Conventional bio-drying pH

Thermal-assisted bio-drying pH and thermal drying

Thermal drying temperature

Thermal assisted bio-drying temperature

Thermal-assisted bio-drying oxygen uptake rate

Thermal drying oxygen uptake rate

FIGURE 3.3
Bio-drying characteristics diagram. Adapted from Liu et al. (2016).

parameters as a different heat source from drying technologies can create a transition period in the middle of the drying characteristics and increase the diffusivities. Table 3.1 shows effective diffusivities of different food waste using different drying technologies. It consists of sawdust, stalk food, fruit, vegetables, pits, and cooked food waste. The effective diffusivity values ranged from 10^{-16} to $10^{-7}\,\mathrm{m^2/s}$. The high and low effective diffusivity values are also dependent on the food waste characteristics like oil contents and thickness. In this chapter, different drying technologies used in the drying of food waste are discussed and potential areas to explore are revealed.

3.3 Combined Drying

Combined drying refers to the application of a combination of two or more drying systems to achieve drying objectives. In some situations, combined drying is necessary to compensate for the limitations of other drying methods. For example, using a solar drying method alone throughout the year may not be suitable due to inconsistent weather conditions and the change of seasons might cause inadequate drying as the final moisture content remains high. Another example where combined drying is needed is when dealing with nutritional properties that easily deteriorate like saffron residues. A high-end drying technology like freeze drying (FD) can be used as well to preserve nutritional properties.

Drying may also be combined with pre- or post-treatment to achieve products of a specific quality. For instance, drying can be combined with shredding for a faster drying process (Ma et al., 2018) or combined with grinding to produce fine-size materials such as flour (Bas-Bellver et al., 2020). Some materials required blanching as pre-treatment to minimise the loss of original

TABLE 3.1

Examples of Effective Diffusivities of Food Waste Dried Using Different Drying Technologies

Food Waste	Effective Diffusivity Values, m²/s	Remarks	References
Sawdust	Oxidative atmosphere 4.87×10^{-13} to 1.49×10^{-9} Inert atmosphere 1.57×10^{-13} to 1.20×10^{-9}	The lowest D_{eff} values were obtained when the sawdust dried under an inert atmosphere.	Fernandez et al. (2018)
Stalk	Oxidative atmosphere 6.67×10^{-16} to 8.29×10^{-8} Inert atmosphere 1.60×10^{-17} to 1.72×10^{-9}		
Food, fruit, and vegetable residues	Air velocity: 2 m/s 3.65×10^{-8} (70°C) 1.83×10^{-8} (60°C) 9.13×10^{-9} (50°C) Air velocity: 1.5 m/s 7.3×10^{-9} (70°C) 6.39×10^{-9} (60°C) 4.56×10^{-9} (50°C) Air velocity: 1 m/s 6.39×10^{-9} (70°C) 3.65×10^{-9} (60°C) 2.74×10^{-9} (50°C)	Cabinet dryer with a conventional tray. The particle sizes were less than 20 mm. Thickness of samples was set at 3.0 mm and dried until the moisture ratio of 0.2. The initial weight and final moisture content are not disclosed. The lowest energy consumption was set at 70°C and 2 m/s and the highest energy consumption was at 50°C and 1 m/s. The total energy required for drying food waste ranged from 59.41 to 119.62 kW h. Mathematical modelling – No.	Khaloahmadi et al. (2021)
Peach and plum pits from canneries and jam factories, olive pits from oil industry, marc and stalk from wineries and pine sawdust from sawmills		The sample size ranged from 0.10 to 0.21 mm. The D_{eff} values increase with respect to the obtained values under inert atmosphere.	

(Continued)

TABLE 3.1 (*Continued*)
Examples of Effective Diffusivities of Food Waste Dried Using Different Drying Technologies

Food Waste	Effective Diffusivity Values, m²/s	Remarks	References
Plum pits	Oxidative atmosphere 4.10×10^{-9} to 2.18×10^{-7} Inert atmosphere 1.86×10^{-11} to 5.24×10^{-9}	Oxidation reactions occurred during the drying stage to release volatile compounds. Under an oxidative atmosphere, the carboxyl and carbonyl groups increase gradually with the oxidation temperature up to 423 K and not under a nitrogen atmosphere.	Fernandez et al. (2018)
Olive pits	Oxidative atmosphere 4.41×10^{-10} to 1.33×10^{-7} Inert atmosphere 3.91×10^{-12} to 8.72×10^{-9}	The highest D_{eff} values were predicted when the olive pits were dried under oxidative atmosphere.	Fernandez et al. (2018)
Peach pits	Oxidative atmosphere 1.25×10^{-9} to 1.54×10^{-6} Inert atmosphere 3.90×10^{-10} to 9.22×10^{-7}		Fernandez et al. (2018)
Marc	Oxidative atmosphere 1.01×10^{-12} to 2.82×10^{-9} Inert atmosphere 8.64×10^{-16} to 2.07×10^{-9}		Fernandez et al. (2018)

(*Continued*)

TABLE 3.1 (*Continued*)

Examples of Effective Diffusivities of Food Waste Dried Using Different Drying Technologies

Food Waste	Effective Diffusivity Values, m²/s	Remarks	References
Cooked food (rice, bread, noodles, meat, spaghetti, beans, mung bean, and peas), vegetables or uncooked food waste (onion peels, potato peels, coriander stems, mint stems, green peppers, tomatoes, eggplant peels, cucumber peels), and fruit waste (melon peel, watermelon peel, apples, apricots, etc.)	Solar drying Non-optimised 6.10×10^{-10} (day 1) 8.92×10^{-10} (day 2) 7.20×10^{-10} (day 3) Optimised shrinkage not considered 3.30×10^{-10} (day 1) 3.75×10^{-10} (day 2) 4.12×10^{-10} (day 3) Optimised shrinkage considered 2.33×10^{-10} (day 1) 2.56×10^{-10} (day 2) 2.81×10^{-10} (day 3) Lab. batch Non-optimised 5.69×10^{-10} Optimised shrinkage not considered 2.93×10^{-10} Optimised shrinkage considered 1.88×10^{-10}	500 g food waste. Solar drying temperature ranged from 28°C to 37°C, atmospheric pressure: 82 kPa, relative humidity ranged from 12% to 35%, and sun radiation ranged from 50 to $1100 \, W/m^2$. Convective air-drying temperature: 55°C, air flow rate: 25 L/s, and initial material layer thickness was the same as in the solar dryer (5 mm). Water activity of food waste: 0.3 and final moisture content <6%. For optimised shrinkage calculation, the linear decrease of half of the layer thickness was considered. Solar dryers are more sustainable.	Noori et al. (2022)

(*Continued*)

TABLE 3.1 (*Continued*)
Examples of Effective Diffusivities of Food Waste Dried Using Different Drying Technologies

Food Waste	Effective Diffusivity Values, m²/s	Remarks	References
Rambutan seed	Oven drying 9.74×10^{-8} m²/s (40°C) 1.76×10^{-7} m²/s (60°C) Microwave drying 3.13×10^{-7} (50 W/g) 1.59×10^{-7} (200 W/g)	5.0 g of samples. Microwave drying at lower microwave power can shorten the drying duration and save more energy.	Ahmad et al. (2017)
Nectarine pomace	0.279–3.15×10^{-9}	30°C–70°C for hot air, and 50°C–70°C for hot plate, with 10°C step difference. 50°C–70°C, or without heating (WH). 5, 7, and 10 mm. 82.5, 56.8, and 45.5 g for the sample thicknesses of 10, 7, and 5 mm, respectively.	Milanovic et al. (2021)
Olive mill waste	TEST A 6.20×10^{-8} (264°C, 4 m/s) TEST B 1.917×10^{-8} (100°C, 4 m/s) TEST C 9.85×10^{-8} (425°C, 4 m/s) TEST D 4.45×10^{-8} (263°C, 1 m/s) TEST E 9.59×10^{-8} (263°C, 7 m/s) TEST F 4.77×10^{-8} (181°C, 5.5 m/s) TEST G 3.40×10^{-8} (181°C, 2.5 m/s) TEST H 10.02×10^{-8} (344°C, 5.5 m/s) TEST I 8.28×10^{-8} (344°C, 2.5 m/s)	61% (w.b.). EMC: 8.5% (w.b.). Rotary drying Temperature ranged from 100°C to 425°C and drying air velocity ranges between 1.0 and 7.0 m/s. Tests A, C, D, E, H, and I experimented with the combustion phenomenon. Combustion was reached in tests A, D, E, and I at EMC. The C and H samples started to burn at moisture content of 23% and 30%. For industry-scale rotary drying, it is recommended to dry at low temperature and air flow as the difference is less than 20%. However, the C and H settings should not be considered as the samples are not able to reach the EMC value.	Gómez-de la Cruz et al. (2020)

(Continued)

TABLE 3.1 (*Continued*)

Examples of Effective Diffusivities of Food Waste Dried Using Different Drying Technologies

Food Waste	Effective Diffusivity Values, m²/s	Remarks	References
Olive pomace waste	Sample thicknesses considered are 0.5, 1.0, and 1.5 cm. Kinetic measurements are carried out for three temperatures (40°C, 60°C, and 80°C) and two drying air flow rates (0.042 and 0.083 m³/s) 9.30×10^{-8} (40°C, 0.042 m³/s, 1.5 cm) 9.01×10^{-8} (40°C, 0.083 m³/s, 1.5 cm) 20.4×10^{-8} (60°C, 0.042 m³/s, 1.5 cm) 20.3×10^{-8} (60°C, 0.083 m³/s, 1.5 cm) 34.7×10^{-8} (80°C, 0.042 m³/s, 1.5 cm) 34.1×10^{-8} (80°C, 0.083 m³/s, 1.5 cm) 7.61×10^{-8} (40°C, 0.042 m³/s, 1.0 cm) 7.76×10^{-8} (40°C, 0.083 m³/s, 1.0 cm)	Solar convective dryer operating in forced convection.	Koukouch et al. (2017)

(*Continued*)

TABLE 3.1 (*Continued*)

Examples of Effective Diffusivities of Food Waste Dried Using Different Drying Technologies

Food Waste	Effective Diffusivity Values, m²/s	Remarks	References
	11.5×10^{-8} (60°C, 0.042 m³/s, 1.0 cm)		
	11.5×10^{-8} (60°C, 0.083 m³/s, 1.0 cm)		
	17.9×10^{-8} (80°C, 0.042 m³/s, 1.0 cm)		
	18.4×10^{-8} (80°C, 0.083 m³/s, 1.0 cm)		
	2.48×10^{-8} (40°C, 0.042 m³/s, 0.5 cm)		
	2.53×10^{-8} (40°C, 0.083 m³/s, 0.5 cm)		
	3.86×10^{-8} (60°C, 0.042 m³/s, 0.5 cm)		
	3.89×10^{-8} (60°C, 0.083 m³/s, 0.5 cm)		
	7.17×10^{-8} (80°C, 0.042 m³/s, 0.5 cm)		
	7.22×10^{-8} (80°C, 0.083 m³/s, 0.5 cm)		

(*Continued*)

TABLE 3.1 (*Continued*)

Examples of Effective Diffusivities of Food Waste Dried Using Different Drying Technologies

Food Waste	Effective Diffusivity Values, m²/s	Remarks	References
Olive pomace	27.5 cm × 18.5 cm × 1.2 cm Thickness, 6 cm 60 1.84×10^{-7} 70 3.03×10^{-7} 80 3.42×10^{-7} Thickness, 9 cm 60 2.17×10^{-7} 70 2.98×10^{-7} 80 3.67×10^{-7} Thickness, 12 cm 60 2.18×10^{-7} 70 3.22×10^{-7} 80 3.94×10^{-7}	Samples with small particle sizes are packed to obtain an almost non-porous structure. However, large particles resulted in a porous structure and the rate of removal of water in such a solid system is easier than the non-porous systems. Large particles are mainly from the pits of the olive, and they contain higher amounts of cellulose compared to the smaller particles. Small particles are from olive fruit itself and contain more oil than pits pomace.	

texture, colour, and flavour, by immersing them in boiling water for a short time followed by immersion in cold water (Duarte et al., 2017). Pre- or post-treatment of material after drying utilising chemicals is employed to acquire targeted properties, for instance, to improve the adsorptive properties of the material (Plazzotta et al., 2018). Several examples of combined drying methods used for drying various food wastes are summarised in Table 3.2.

Kobayashi et al. (2019) and Ma et al. (2018) studied the use of thermally assisted bio-drying. Conventionally, bio-drying depends solely on high microbial activity to generate heat and subsequently reduce the water content in materials through evaporation. However, in thermally assisted drying, the drying process was expedited. The introduction of char as additional material during bio-drying helped to increase the drying rate of the material and increased the decomposition rate. In addition, staged heating acclimation during thermally assisted bio-drying results in excellent thermophilic inoculum with high metabolic activity and microbial consortia.

Mutlu et al. (2021) studied the drying of brewery spent grains (BSG) at 75°C and at air flow 180,000 m^3/h utilising solar drying integrated with a biomass boiler. The biomass boiler was introduced to the solar drying system for two reasons: (1) to enable efficient drying at fluctuating weather conditions and (2) for continuous drying during nights. It is concluded that controlled inlet air and BSG flow rates are crucial to balance the solar wall heat production fluctuations at varying weather conditions. In a similar study by Wallin et al. (2020), it is concluded that the proposed system is feasible only if the electricity cost is <5 times higher than the biofuel cost.

The comparison between the final products of drying combined with grinding of iceberg salad fresh-cut processing waste in hot-air drying (HAD), FD, and supercritical-CO$_2$-drying (S CO$_2$) was studied by Plazzotta et al. (2018). Results showed that HAD produced flour rich in fibre (>260 g/kg) and polyphenols (3.05 mg GAE/g$_{dw}$) with good anti-oxidant activity. Freeze-drying was able to preserve the vegetable structure and colour better than HAD at low temperature. The freeze-dried sample retained some polyphenols in the dried product but lower than hot-air-dried sample (1.23 mg GAE/g$_{dw}$). Supercritical-CO$_2$-drying with ethanol as co-solvent followed by grinding produced some flour with an excellent ability to absorb water and oil.

Calín-Sánchez et al. (2020) studied the comparison between HAD combined with extrusion and MD combined with extrusion for treating the bilberry press cake extrudates. Results showed that the extrusion process accelerated the drying process. Further to this, a 40% time reduction can be achieved with microwave drying (MD) extrusion, compared with HAD extrusion, without compromising the product quality. Samples dried via HAD and MD have similar retention of total phenolic compounds (TPC) and physical characteristics.

Ko et al. (2021) and Rodriguez et al. (2019) studied the combination of HAD with other drying systems. Rotten citrus fruits were dried to compost using

TABLE 3.2

Examples of Combined Drying Methods Used for Drying of Various Food Waste

Food Waste and Category	Drying Method	Research Findings	Potential Venture	References
Category A: Starch- and Fibre-Rich				
Dog food (as simulated organic waste)	Thermally assisted bio-drying with sludge char addition.	Sludge char increased the drying rate of dog food during bio-drying and increased the decomposition rate.	A lab-scale experiment. Adding char to organic affected the effective diffusivity, which could have an impact on the dog food product quality. It has the potential to develop as high-quality compost.	Kobayashi et al. (2019)
Mixed food waste (rice, vegetable, and meat)	Thermal-assisted bio-drying. Mode 1: stepwise increase: 35°C–50°C. With the increment of 5°C. Mode 2: 50°C.	Staged heating acclimation results in excellent thermophilic inoculum with high metabolic activity and microbial consortia.	A lab-scale experiment. The drying effects after 4 days are equivalent to conventional bio-drying for 20 days (about 3 weeks). It has the potential to be converted to industrial-/large-scale drying.	Ma et al. (2018)
Saffron floral bio-residues	FD: 24, 48, and 72 hours; HAD: 50°C, 70°C, 90°C, 110°C and 125°C.	Freeze drying is required to stabilise saffron floral bio-residues because they are easily deteriorated. Anthocyanins and flavanols degraded at 110°C and 125°C. Recommended drying condition for best colour, flavanols, and anthocyanins content: 90°C with air flows 2, 4, or 6m/s.	A lab-scale experiment and sample mass are only 100g. The recommended operating conditions preserved the colour and bioactive compounds of food waste. The HAD drying has the potential to be converted to industrial-/large-scale drying.	Serrano-Díaz et al. (2013)

(Continued)

TABLE 3.2 (*Continued*)
Examples of Combined Drying Methods Used for Drying of Various Food Waste

Food Waste and Category	Drying Method	Research Findings	Potential Venture	References
Brewery spent grains (BSG)	SD combined with biomass boiler.	Controlled inlet air and BSG flow rates are crucial to balance the solar wall heat production fluctuations at varying weather conditions.	A large-scale simulation-based research has two heating stages with an area of 2,500 m² (about the area of a large mansion), a double-pass rotary dryer, and a multi-fuel biomass boiler with a thermal output of 950 kW. However, the exact amount of sample dried per badge was not addressed which has an impact on the efficiency of the drying process proposed.	Mutlu et al. (2021)
	SD combined with biomass boiler.	The proposed system is feasible only if the electricity cost is <5 times higher than the biofuel cost.	This is another simulation-based research paper but focuses on energy consumption. The exact amount of sample dried per badge was not addressed, which has an impact on the efficiency of the drying process proposed.	Wallin et al. (2020)
Category B: Fruit and Vegetable Waste				
Bilberry press cake extrudate	HAD combined with extrusion/MD with extrusion.	40% time reduction with microwave drying. Samples dried via HAD and MD have similar retention of total phenolics and physical characteristics.	A lab-scale experiment and the number of samples produced per batch were not discussed which has an impact to scale up. It has the potential to be converted to pilot-scale drying.	Höglund et al. (2018)
Iceberg salad fresh-cut processing waste	HAD and grinding.	Product: flour rich in fibre and polyphenols, and good anti-oxidant activity.	A lab-scale experiment and sample and samples are 1 kg per batch. Dried waste has the potential to develop as functional food products. It has the potential to be converted to pilot-scale drying.	Plazzotta et al. (2018)

(*Continued*)

TABLE 3.2 (*Continued*)

Examples of Combined Drying Methods Used for Drying of Various Food Waste

Food Waste and Category	Drying Method	Research Findings	Potential Venture	References
	FD and grinding.	FD sample retained some polyphenols in dried product but lower than hot-air-dried sample. Freeze drying preserved vegetable structure and colour.	A lab-scale experiment and sample and samples are 1 kg per batch.	
	SCO_2 (ethanol as co-solvent) and grinding.	SCO_2 with ethanol as co-solvent followed by grinding produced flour with excellent ability to absorb water and oil.	A lab-scale experiment and sample and samples are 5 g per batch. It has the potential to develop as bulking agents or oil absorbers (to absorb oil spills, edible oils).	
Rotten citrus fruit	MD plasma and HAD (liquefied petroleum gas as heat source) for drying and composting.	An air velocity of 5 m/s was required to obtain a suitable temperature for drying rotten citrus fruits. The reactor temperature dropped with increasing air flow rate.	Pilot-scale research combined with simulation. The product quality is not assessed in this research. By taking account of the time, space, and energy saving during drying, it has the potential to develop as a compost.	Ko et al. (2021)
Nopal pads/ cactus pear peel (*Opuntia ficus-indica*)	OD as a pre-treatment before HAD.	The combined drying is more efficient in drying sample compared to HAD alone. OD process could result in unacceptable nopal pads taste due to a higher solid gain.	A lab-scale experiment but samples dried per batch are not mentioned. This is an easy and cost-saving method for preserving nutritional qualities of nopal pads/peels as ingredients in functional food. It has the potential to be applied to other wastes that fall in Category B.	Rodriguez et al. (2019)

(Continued)

TABLE 3.2 (Continued)

Examples of Combined Drying Methods Used for Drying of Various Food Waste

Food Waste and Category	Drying Method	Research Findings	Potential Venture	References
Nopal pads/Cactus pear peel (*Opuntia ficus-indica*)	Swell-drying coupling instant controlled pressure drop with hot-air drying.	The process obtained dried product with high crispness and improved structure, nutritional content, and relative expansion ratio (up to 21 times).	A lab-scale experiment but samples dried per batch are not mentioned. Setting it at low pressure can improve the product quality but cost analysis is not performed in this study, which could have an impact on the feasibility of scale it up. It has the potential to be developed as snacks.	Namir et al. (2017)
Broccoli stalk slices	OD with MD-assisted HAD.	The minimum energy input required to achieve a final moisture content of 10% is about 2.46 MJ/kg of evaporated water. It can be reduced by 68% with a return on investment (ROI) of more than 30% in 2 years.	This is a techno-economical aspect of Integrated Microwave Drying research paper. A similar model can be used to predict other large-scale food waste drying from Categories A to C.	Salina MD Salim et al. (2016)
Apple peels	Solar-electrical forced HAD. Radiation: 400–900 W/m²; temperature: 50°C–80°C; relative humidity: 12%–38%; duration: 110–280 minutes.	Lowest total energy consumed and maximum energy efficiency at a higher temperature and a lower airflow. In addition, the bioactive compounds degrade with the increased drying temperature.	A pilot-scale experiment. It has the potential to be converted to large-scale processes. In addition, food waste falls into Category B and can attempt to use this drying method.	Moussaoui et al. (2021)

(Continued)

TABLE 3.2 (*Continued*)

Examples of Combined Drying Methods Used for Drying of Various Food Waste

Food Waste and Category	Drying Method	Research Findings	Potential Venture	References
Category C: Others				
Salmon bone	Boiling by gas retort followed by HAD. HAD: 140°C (45 minutes), temperature (duration): 160°C (30–45 minutes), 180°C (15–45 minutes).	Recommended HAD at 180°C for 15 minutes to produce salmon bone snack with crispness (84 count peaks), hardness (13.16 N), moisture content (2.96%), and calcium (19.23 g calcium per 100 g).	A lab-scale experiment. The amount of sample dried was not mentioned which has an impact on the quality of the product and drying duration. Reducing the cut length has the potential to reduce the drying duration and temperature. The relation between drying conditions and seasoning process of the product is yet to be studied.	Hirunrattana and Limpisophon (2019)

microwave plasma and HAD. For HAD, liquefied petroleum gas was used as a heat source. The study found that an air velocity of 5 m/s was required to obtain a suitable temperature for drying rotten citrus fruits. In a study by Rodriguez et al. (2019), nopal pads (*Opuntia ficus-indica*) underwent osmotic dehydration (OD) as a pre-treatment before HAD. Results showed that combined drying is more efficient in drying samples compared to HAD alone. However, the OD process could result in an unacceptable nopal pad taste due to a higher solid gain.

Based on the combined drying approaches discussed, it can be concluded that the introduction of MD into a combined drying system generally results in faster drying, without changing the characteristics of the product. However, there is no specific energy consumption and cost analysis to determine the feasibility of pilot or large-scale processing. In addition, particle size has an impact on the drying duration but is not considered part of the research scope in food waste drying. The impact of changing the particles significantly impacts the effective diffusivity values compared to increasing the power, temperature, or pressure of the system (Serrano-Díaz et al., 2013). FD is not suitable for food waste drying in the industry even though it can be used to preserve important nutritional qualities while combining high temperature into bio-drying results in improved drying rate and decomposition rate due to its cost and operating duration.

3.4 Hot-Air Drying

Hot-air drying (HAD) is a common method for drying. It is important in any HAD process to optimise the drying conditions to ensure the bioactive compounds are preserved as much as possible, especially for the application as a nutritional supplement. Table 3.3 shows some examples of HAD methods used for drying various types of food waste.

Scalcon et al. (2018) studied the use of HAD at 80°C, 110°C, and 140°C and mass flow rate of 0.5 kg/min for drying digested sludge derived from gelatine processing. They concluded that the sludge's apparent density and porosity vary to drying rates and critical moisture content. This study also confirmed the shrinkage during drying, which was evidenced by the decreasing thickness, volume dimensions, and altered surface morphology. However, the porosity content of dried sludge is independent of drying temperatures.

The HAD method was also used to dry various by-products of olive oil processing such as olive waste cake (Pasten et al., 2019) and pitted olive pomace (Sinrod et al., 2019). Ahmad-Qasem et al. (2014) studied the influence of HAD and freeze drying on the bio-accessibility of bioactive polyphenols in the olive leaf extract and concluded that the bio-accessibility in samples dried using HAD and freeze drying was comparable. Meanwhile,

TABLE 3.3

Examples of HAD Methods Used for Drying Various Types of Food Waste

Food Waste and Category	Drying Conditions	Key Findings	Potential Venture	References
Category A: Starch- and Fibre-Rich				
Gelatin sludge	Temperature: 80°C, 110°C, and 140°C; mass flow rate: 0.5 kg/min	Sludge apparent density and porosity affected the drying rates and critical moisture content. Shrinkage during drying was confirmed with decreasing thickness, surface morphology, and volume dimensions. Porosity content is independent of drying temperatures.	A lab-scale experiment. The drying temperature ranged from 80°C to 140°C. The amount of sludge per batch of drying, energy consumption, and retained nutrients are not specified, which has an impact due to drying.	Scalcon et al. (2018)
Brewery spent grains (BSG)	Temperature: 40°C (48 hours)	Moisture content reduction: 5.44–5.57 g/100g; water activity reduction: 0.19–0.20. This drying method retained a higher number of bioactive compounds than HAD and at the same time lowered the moisture content and water activity.	A lab-scale experiment. It was reported that impingement drying is better than HAD. It has the potential to be converted to pilot-scale drying. However, energy consumption and cost analysis which have an impact on the feasibility of the drying method were not assessed in this study.	Shih et al. (2020)
Brewers' spent grains (BSG)	Temperature: 60°C, 65°C, and 70°C; air flow 0.05 m/s	Produced dried chips at 50 minutes drying time, 65°C. Product quality is lower compared to MVD and FD.	A lab-scale experiment. It was compared with MVD and FD. The drying duration of HAD is longer compared to others but combining with MVD for pilot- or large-scale processing is yet to be investigated.	Pratap Singh et al. (2020)
Coffee pulp (*Coffea canephora*)	Temperature: 70°C (7 hours 45 minutes); 90°C (4 hours 45 minutes); 110°C (3 hours 30 minutes)	Increased air circulation during drying causes exposure to oxygen and high temperature, resulting in oxidation of some phenolic compounds in dried products. In this study, vacuum drying is recommended as it preserved more bioactive compounds.	A lab-scale experiment. The drying temperature recommended for vacuum drying is as high as 110°C more than 4 hours. However, energy consumption is not assessed in this paper, which has impacted the decision on a bigger scale process.	Kieu Tran et al. (2020)

(Continued)

TABLE 3.3 (Continued)

Examples of HAD Methods Used for Drying Various Types of Food Waste

Food Waste and Category	Drying Conditions	Key Findings	Potential Venture	References
Palletised mixed wastes (straw and broken kernels of wheat, bread, date palm, grape, pomegranate, and potato and soybean meal)	44.85°C, 59.85°C, 74.85°C	Drying at high temperature caused some cracks in the samples. Crude fat and neutral detergent fibre have an impact on mechanical characteristics.	A lab-scale experiment. This research project has the potential to scale up after assessing the energy consumption, retained nutrients are not specified, which has an impact due to drying.	Ghasemi et al. (2018)
Wheat kernel (from durum and einkorn wheat)	Temperature: 50°C, 105°C or 130°C for 90 minutes	HAD results in high extractability of alkylresorcinols and a decrease in total soluble phenolic compounds after drying process. At 130°C, an increase of long chains and a decrease of short chains were observed.	A lab-scale experiment. The choice of genotype has an impact on the overall process. The evaluation of eco-friendly extraction process to couple with heat treatment pre-processing needs to be further assessed.	Ciccoritti et al. (2021)
Category B: Fruit and Vegetable Waste				
Carrot pomace	Temperature: 60°C–75°C, air flow: 0.5, 0.7 and 1.0m/s	Increase in air flow velocity resulted in an increase of effective diffusivity and a decrease in drying time.	A lab-scale experiment. It is only focused on drying characteristics and drying kinetics of carrot pomace. Retained nutrients or bioactive compounds are not assessed, which has an impact due to the drying operating conditions.	Kumar et al. (2011)

(Continued)

TABLE 3.3 (*Continued*)

Examples of HAD Methods Used for Drying Various Types of Food Waste

Food Waste and Category	Drying Conditions	Key Findings	Potential Venture	References
Cacao pod husks (*Theobroma cacao* L.)	Temperature: 60°C; air flow: 0.1 m/s for 24 hours	MD and FD are better than HAD in retaining total phenolic compounds, flavonoid content, and anti-oxidant activity.	A lab-scale experiment. MD is a potential drying method compared to conventional HAD. It is important to know how samples dry per batch during MD. In addition, energy consumption is an important parameter that has an impact on the drying duration and economic parameters for pilot- or large-scale MD. It should be compared with FD and HAD as well.	Valadez-Carmona et al. (2017)
Kyoho grape seeds (*Vitis labruscana*)	Temperature: 30°C–60°C; relative humidity: 70%; sample load density 1.04 kg/m²; air flow: 1.0 m/s	Drying temperature affects the physicochemical properties of dried seeds except for colour and oil-holding capacity. The drying rate decreased with moisture contents and drying times.	A lab-scale experiment that focused on drying characteristics and kinetics and physicochemical properties. The physicochemical properties should be considered for other food wastes to fall into seed type Category B waste. Energy consumption is an important parameter that has an impact on the drying duration and economic parameters should be assessed before attempting pilot- or large-scale drying.	Sridhar and Charles (2020)
Kiwifruit waste (leather)	Temperature: 70°C, air flow 2 m/s, air renewal fee of 50%	Kiwifruit leather-based snacks prepared by HAD contained lower total phenolics, vitamin C, and flavonoids compared to FD dried snacks.	A lab-scale experiment. Drying set at different temperatures which have an impact on the bioactive compounds is not assessed. In addition, energy consumption can be considered before attempting pilot- or large-scale drying.	Tylewicz et al. (2020)

(Continued)

TABLE 3.3 (*Continued*)

Examples of HAD Methods Used for Drying Various Types of Food Waste

Food Waste and Category	Drying Conditions	Key Findings	Potential Venture	References
Mango peel	Temperature: 60°C–80°C	Separation of puree from fresh industrial mango peel waste before processing into dried mango peel yielded the maximal number of reusable by-products without affecting dried mango peel functionality.	A lab-scale experiment. This is suitable for pilot- and large-scale processing. In addition, it has the potential to integrate with other peels in Category B with mango peels that could be considered for composting or energy recovery, optionally after the recovery of adhering pulp.	Nagel et al. (2014)
Mango peel and kernel	Temperature: 60°C, air circulation, 4 hours	HAD yielded product with lower total phenolics, carotenoids, and anti-oxidant properties compared with freeze drying.	As food ingredient and value-added products (for the kernel: mango kernel oil, mango kernel butter, mango kernel flour, and biofilms).	Sogi et al. (2013)
Mangosteen rind	Temperature: 60°C, 80°C and 100°C	The optimum drying temperature was 60°C, giving products with the highest sensory acceptability score. Recommended to add biopolymers.	A lab-scale experiment. Other drying methods or combined drying methods can be considered to be used in this project which has an impact on drying duration, product quality, and sensory acceptability score.	Sothornvit (2012)
Nectarine pomace	Conductive-convective drying; 30°C–70°C; air flow: 1.5m/s	Hot air has a significant role at initial stages of drying but as drying prolongs, the effect decreases and has negative impact on drying times.	A lab-scale experiment that focused on drying characteristics and kinetics of nectarine pomace. Retained nutrients or bioactive compounds are not assessed, which has an impact due to drying operating conditions.	Milanovic et al. (2021)

(Continued)

TABLE 3.3 (*Continued*)

Examples of HAD Methods Used for Drying Various Types of Food Waste

Food Waste and Category	Drying Conditions	Key Findings	Potential Venture	References
Pitted olive pomace	80°C, 130 min, 24.5 g/m² load density	Drying decreased the anti-oxidant capacity. HAD caused the most thermal damage to olive pomace compared to drum drying and FD.	It is between lab- and pilot-scale experiments. There is a potential to optimise rotational drum operating conditions and to retain higher bioactive compounds. In addition, energy consumption per batch of product dried using different dryers should be assessed.	Sinrod et al. (2019)
Olive waste cake	Temperature: 40°C–90°C; air flow: 2.0 m/s	Fatty acids were found in high concentration at low temperatures. The anti-oxidant capacity decreased with drying.	A lab-scale experiment. Important retained bioactive compounds were assessed in this study. Further to this, energy consumption is an important parameter that has an impact on drying duration and economic parameters should be assessed before attempting pilot- or large-scale drying.	Pasten et al. (2019)
	50°C–90°C	The anti-oxidant activity decreased with increasing HAD temperatures. Vitamin E content increased at temperature 50°C–90°C.	Same as above.	Uribe et al. (2013)
Papaya tissue (pulp or peel)	Temperature: 40°C; air flow: 2.0 m/s	The product of HAD has significantly higher anti-oxidant capacity compared to MD dried sample.	A lab-scale experiment. Energy consumption is an important parameter that has an impact on the drying duration and economic parameters should be assessed before attempting pilot- or large-scale drying.	Nieto-Calvache et al. (2019)
Persimmon slices (overripe, with peels)	Temperature: 40°C and 60°C	HAD is a suitable method for producing dehydrated persimmon snacks. Applying HAD could avoid the use of a de-astringency treatment for overripe persimmon.	A lab-scale experiment. Important retained bioactive compounds were assessed in this study before scaling up to pilot and large scale.	González et al. (2021)

(Continued)

TABLE 3.3 (*Continued*)

Examples of HAD Methods Used for Drying Various Types of Food Waste

Food Waste and Category	Drying Conditions	Key Findings	Potential Venture	References
Shiitake stipes (*Lentinula edodes*)	Temperature: 70°C for 12 hours, air flow of 2.0 m/s	HAD favours the release of free amino acids (FAA) and resulted in samples with high equivalent umami concentration (EUC).	A lab-scale experiment. It is identified as a flavour enhancer for low sodium foods and has the potential to scale up to pilot scale.	França et al. (2022)
Watermelon rind (*Citrullus lanatus*)	Temperature: 50°C	The water absorption capacity significantly decreased, while the water solubility and bulk density increased. A lower number of tannins, alkaloids, and saponin were found in HAD product compared to FD.	A lab-scale experiment. The studied operating temperature can be further increased to reduce the drying duration. Further to this, assess the bioactive compounds retained.	Mohan et al. (2016)
Yellow passion fruit peel (*Passiflora edulis var. Flavicarpa*)	Temperature: 50°C, 55°C or 60°C; air flow 2.4, 2.7 and 3.0 m/s	The studied drying conditions are appropriate for drying the sample and do not affect the physicochemical properties of the dried peel. The shortest drying time (2.5 hours) was achieved at a temperature of 60°C and an airflow rate 2.7 m/s.	A pilot-scale experiment. However, the actual amount dried per batch is not specified. Focus on drying characteristics and kinetics only. The energy consumption and retained bioactive compounds are important parameters that have an impact on the economic parameters and should be assessed before attempting large-scale drying.	Duarte et al. (2017)

olive waste cake HAD studied by Uribe et al. (2013) found that the anti-oxidant activity of the olive waste cake decreased with increasing HAD temperatures, particularly at 90°C. At the same time, Vitamin E content increased after the olive waste cake dried at 50°C–90°C. In addition, Pasten et al. (2019) showed that the abundance of fatty acid types varies with drying temperature. Low drying temperatures (40°C–50°C) yielded waste cake with high content of oleic acid, up to 63.9% of the total fatty acids. HAD at 60°C favours the abundance of linolenic acid, which are 12.7 times higher compared to the fresh olive waste cake. Temperatures above 60°C, however, reduce the amount of palmitic and oleic acids in waste cake, although an adequate number of bioactive compounds were retained. Thus, employing optimum drying temperature is important to maximise the utilisation of olive waste cake for nutraceuticals, food supplements, and cosmetic applications.

High drying temperature during HAD influences the chemical properties of dried food waste. HAD at high temperature could cause a decrease in phenolics for some food waste samples like coffee pulp and pitted olive pomace (Ciccoritti et al., 2021; Kieu Tran et al., 2020) while increasing the phenolics for some other food wastes like cacao pod husk (Valadez-Carmona et al., 2017). Increased HAD temperature and air circulation during drying of coffee pulp at 110°C for 3.5 hours caused exposure to oxygen and high temperature, resulting in oxidation of some phenolic compounds in dried product (Kieu Tran et al., 2020). The HAD of wheat kernel results in high extractability of alkylresorcinols and a decrease in total soluble phenolics, while at higher temperature (130°C), an increase of long chains and a decrease of short chains were observed (Ciccoritti et al., 2021). The decrease in phenolics contributes to the decrease in anti-oxidant activity.

It is also interesting to note the differences between the anti-oxidant activity of HAD products in comparison with other drying methods. For instance, HAD results in products with higher anti-oxidant activity compared to microwave-dried products from papaya tissue (Nieto-Calvache et al., 2019). Further to this, the anti-oxidant properties were lower compared to freeze-dried products from mango peels (Sridhar and Charles, 2020).

High drying temperatures also cause breakage or cracks in fibrous food waste samples. During HAD of palletised mixed wastes consisting of straw and broken kernels of wheat, bread, date palm, grape, pomegranate, and potato and soybean meal, cracks were observed on dried samples even at moderate temperature (75°C) (Shih et al., 2020). In the case of pitted olive pomace drying, HAD caused the most thermal damage to olive pomace compared to drum drying and freeze drying (Sinrod et al., 2019). Meanwhile, in Kyoho grape seed drying, temperature affects the physicochemical properties of dried seeds except for colour and oil-holding capacity (Sridhar and Charles, 2020).

3.5 Microwave Drying

In recent years, microwave drying has become one of the most efficient drying methods due to its fast and cost-effective heating process, energy efficiency, and the end-product quality. Microwave drying is suggested as an alternative convective hot-air drying because it enables fast drying, attributed to the direct energy absorption within the product due to dipole rotation and ionic polarisation (Nagel et al., 2017). Microwave drying speeds up the rate of drying by reducing the activation energy at low temperature and vice versa at different drying stages (Liu et al., 2016).

Table 3.4 shows examples of microwave drying methods used for drying various types of food waste. There are several variations of microwave drying, for example, microwave drying, microwave-assisted freeze drying (MFD), microwave-assisted convective drying (MCD), microwave-assisted vacuum drying (MVD), microwave-assisted spouted bed drying (MWSBD), and pulse-spouted microwave-assisted vacuum drying (PSMVD). Basically, MVD is based on vacuum drying technique with microwaves as a source of heat for internal, rapid heating. MVD uses less energy than microwave drying, and the quality of the product dried by MVD is comparable to freeze drying and far superior to vacuum drying (Lin et al., 2021). Meanwhile, in PSMVD, more uniform heating than MVD is achieved via air-spouting of the materials in the microwave dryers. MFD can produce product with high quality and reduce operating costs during food waste drying. However, drying materials with high water content using MFD is ineffective because of low dielectric constant and loss factor of frozen water (Ran et al., 2019).

Microwave drying generally preserved total phenolic compounds (TPC) of food waste better than HAD. However, microwave drying could cause a decrease in some other chemical properties, for example, betacyanin and betaxanthin content (Chew and King, 2019). Preservation of nutritional qualities by microwave drying is also satisfactory, and at times, it is as good as freeze-drying method (Valadez-Carmona et al., 2017). Şahin et al. (2018) concluded that microwave drying is a more suitable method for preserving phenolic compounds in olive leaves during storage, compared with HAD. This is because exposing olive leaf to high temperature during hot-air drying causes a significant loss of phenolic compounds (Valadez-Carmona et al., 2017).

Valadez-Carmona et al. (2017) found that in drying cacao pod husks, a higher level of oleuropein was obtained using microwave drying method compared with HAD. Besides, more anti-oxidant compounds and inactivated polyphenol oxidases were released via microwave drying. In contrast, another study found that HAD-dried papaya peel and pulp had significantly higher anti-oxidant capacity than microwave-dried ones (Nieto-Calvache et al., 2019).

Chew and King (2019) studied the comparison between microwave drying (75, 225, 375, 525, and 750 W) and HAD (condition: 100°C, 4 hours) for

TABLE 3.4

Examples of MD Methods Used for Drying Various Types of Food Waste

Food Waste and Category	Drying Conditions	Key Findings	Potential Venture	References
Category A: Starch- and Fibre-Rich				
Food waste (vegetable leaves and cooked rice)	Power: 750 W and 1,500 W; temperature: 105°C	MD speeds up the rate of combustion by changing the activation energy at different drying stages.	A lab-scale experiment. Important to report the result W/g of sample per batch for future scale-up. In addition, energy consumption and retained bioactive compounds are important parameters that have an impact on the economic parameters and should be assessed before attempting large-scale drying.	Liu et al. (2016)
Food waste-sawdust mixture	Power: 1.0 kW; duration: 0–25 minutes; temperature: 200°C	Time needed to achieve refuse-derived fuel (RDF) standards moisture content (<10%): 15 minutes. The calorific value of dried product increased with more addition of saw dust in mixture.	A lab-scale experiment. The drying process can be further improved to produce better grade solid fuel, miniaturisation of the facility, and fast operation time. Once optimised, it can be tested on a pilot scale.	Choi et al. (2015)
Brewers' spent grains (BSG)	Power: 250 W; vacuum levels ≤0.24 bar	Produced dried chips at 48 minutes drying time and has high drying effectiveness. VMD-derived BSG chips have high overall acceptability.	Refer to Table 3.2, BSG column.	Pratap Singh et al. (2020)
Okara (soy pulp)	Power: 400 W; temperature: 110°C	Lower denatured protein content (% PD) and faster moisture loss compared to RD and FD. Dried okara flour has smooth, rounded structures in a dense matrix.	It is a lab-scale experiment. It is important to report the setting of MD based on the W/g of the sample. For human supply, it is important to report retained bioactive compounds. In addition, energy consumption should be considered as it has an impact on the economic parameters and should be assessed before attempting large-scale drying.	Ostermann-Porcel et al. (2017)

(Continued)

TABLE 3.4 (*Continued*)

Examples of MD Methods Used for Drying Various Types of Food Waste

Food Waste and Category	Drying Conditions	Key Findings	Potential Venture	References
Olive (*Olea europaea*) leaf	Power: 300–500 W; duration: 4–6 minutes; solid mass 1.5–2.5 g	MD is a suitable method for preserving phenolic compounds in olive leaves during storage. RSM-based optimum drying conditions: 2.085 g sample, 459.257 W, 6 minutes.	It is a lab-scale experiment. It is important to report the setting of MD based on the W/g of the sample. Energy consumption and retained bioactive compounds were assessed. It has the potential to attempt pilot and large-scale drying.	Şahin et al. (2018)
Spent coffee grounds briquette	Power: 119W–700 W, with mineral additives (10% sodium chloride/10% sodium sulphate/10% lignite)	Sodium chloride and sodium sulphate improved microwave drying performance, while lignite negatively impacted the microwave drying kinetics of coffee grounds briquette.	It is a lab-scale experiment. It is important to report the setting of MD based on the W/g of the sample.	Fu and Chen (2020)
Category B: Fruit and Vegetable Waste				
Banana peel	Shade drying (7 days) followed by MD at Power: 100 W, 440 W and 1000 W.	Prolonged drying at low input power results in an increase in the final moisture content of the dried product. Drying at 1,000 W produced dried peel faster, with lower final moisture content. Banana peel powder obtained at 1,000 W has the best flow ability.	The amount of sample dried per batch is not specified. It is important to report the setting of MD based on the W/g of the sample. The energy consumption and retained bioactive compounds are important parameters which have an impact on the economic parameters and should be assessed before attempting pilot- or large-scale drying.	Anuar et al. (2019)
Cacao pod husks (*Theobroma cacao L.*)	Power: 595 W for 11.5 minutes; thickness: 1–3 mm	MD released anti-oxidant compounds, inactivated the polyphenol oxidase and maintained the husk micro-structure.	Refer to Table 3.2, cacao pod husks	Valadez-Carmona et al. (2017)

(*Continued*)

TABLE 3.4 (*Continued*)

Examples of MD Methods Used for Drying Various Types of Food Waste

Food Waste and Category	Drying Conditions	Key Findings	Potential Venture	References
Papaya tissue (pulp or peel)	Temperature: 40°C, power: 450 W	MD shortens the processing time while preserving uronic acids content.	Refer to Table 3.2, papaya tissue (pulp or peel).	Nieto-Calvache et al. (2019)
Papaya peel	Power: 250 W (17 minutes), 440 W (10 minutes), and 600 W (8 minutes)	The activation energy value: 46.621 W/g The most suitable input power for thin-layer drying of papaya peel is 600 W.	A lab-scale experiment. The energy consumption and retained bioactive compounds are important parameters that have an impact on the economic parameters and should be assessed before attempting pilot- or large-scale drying.	Manzoor et al. (2019)
Pitaya peel	Power: 75 W, 225 W, 375 W, 525 W, and 750 W	MD resulted in decreased betacyanin and betaxanthin contents. Dried pitaya peel has high rehydration ability.	A lab-scale experiment. The MD method can be further improved to prevent the degradation of betacyanin (BC) and betaxanthin (BX).	Chew and King (2019)
Mango cultivar	Intermittent MD-HAD; power: 800 W, frequency: 2.45 GHz	MD drying enabled a high throughput but yielded a low amount of dried product.	A pilot-scale experiment. This is suitable for Category B food waste for recovery of bioactive compounds and pectin. In addition, the food waste has the potential to be converted into biogas for cogeneration plants.	Nagel et al. (2017)
Orange peels	Power: 334 W, vacuum pressure: 877 mbar	VMD increased the drying rate and effective water diffusivity compared to tray and IR drying. MD preserved quality characteristics and volatile compounds of orange peels.	A lab-scale experiment. It is important to know how many W/kg of sample dry per batch. In addition, energy consumption is an important parameter that has an impact on the drying duration and economic parameters for pilot- or large-scale drying.	Bozkir et al. (2021)

the drying of pitaya peels. Results showed that microwave-dried pitaya peels had higher total phenolic compounds (TPC), comparable antioxidant activity with the fresh peels, and higher rehydration ability than HAD-dried peels. However, a decrease in betacyanin and betaxanthin was observed in pitaya peel after microwave drying (Chew and King, 2019). For papaya peel and pulp, results showed that microwave drying preserved uronic acids better than hot-air convection drying (Sogi et al., 2013).

Microwave drying is also a suitable method to preserve the physical structure of dried food waste. Microwave drying produced dried okara flour with smooth, rounded structures in a dense matrix. Microwave drying also maintained the cacao pod husk micro-structure better compared with HAD (Valadez-Carmona et al., 2017). In addition, papaya pulp undergone microwave drying showed better hydration properties compared to the one dried using HAD (Nieto-Calvache et al., 2019).

In general, higher input power and smaller particle size or thin-layer samples result in shorter drying time of food waste. Choi et al. (2015) studied the drying of food waste-sawdust mixture for the purpose of refuse-derived fuel (RDF) application. They found that time needed to achieve RDF standard moisture content (<10%) was 15 minutes at 1,000 W. Studies on microwave drying of papaya by-product revealed that microwave drying shortens the processing time while preserving uronic acids content (Nieto-Calvache et al., 2019). Manzoor et al. (2019) found that thin-layer microwave drying of papaya peel gave the activation energy value of 46.621 W/g and drying time decreased with increasing input power: 250 W (17 minutes), 440 W (10 minutes), and 600 W (8 minutes). In contrast, longer time was needed for drying thin layers of brewery's spent grain (BSG) via MVD at 250 W, whereby dried BSG chips were produced in 48 minutes (Pratap Singh et al., 2020).

3.6 Solar Drying

In this section, only full solar drying is included for the purpose of analysis and evaluation. Solar drying (SD) in its hybrid form is discussed in the section "Combined drying." Table 3.5 shows examples of SD methods used for drying various types of food waste.

Montero et al. (2011) studied the thin-layer passive solar drying of olive oil processing waste, namely the olive pomace, sludge, and olive mill wastewater. They found that drying time depends on the type of waste. The effective diffusivity values were found to be higher in olive pomace, followed by wastewater and sludge, resulting in longer drying time for sludge. Generally,

TABLE 3.5

Examples of Solar Drying Method Used for Drying Various Types of Food Waste

Food Waste and Category	Drying Conditions	Key Findings	Potential Venture	References
Category A: Starch- and Fibre-Rich				
Annatto grains/annatto waste grains flour	Solar dryer, temperature: $\geq 50°C$	Grain annatto took longer time to reach moisture content ~5% compared to waste grains flour. Effective diffusivity: $6.47 \times 10^{-11} m^2/s$ for annatto grain and $0.69 \times 10^{-11} m^2/s$ for waste grains flour.	High potential to convert into carbohydrates, protein, and dietary fibre food products. However, the product quality was not assessed during solar drying.	Santos et al. (2014)
Olive pomace, sludge and olive mill wastewater	Temperature: 20°C–50°C; operation mode: passive (natural convection) or active (forced convection)	Drying time decreased with increasing temperature and mass flow. Employing the active mode (forced convection) solar drying decreased the drying time.	High potential to convert into fertilisers and human diet. However, the product quality was not assessed. If the product used fuel or compost, a bioactive compound study is not required.	Montero et al. (2011)
Sargassum (golden tides)	SD duration: 2–3 days in sunny weather, 7 days in rainy season	Least expensive method but required large area. Solar drying can cause considerable denaturisation of organic compounds.	High potential to convert into food, fuel, and pharmaceutical products. However, the product quality was not assessed.	Milledge and Harvey(2016)
Category B: Fruit and Vegetable Waste				
Tomato waste	Greenhouse 5-walls solar dryer, inclined roof (36°), 40°C–58°C	Drying time for 4 mm thick tomato waste to afford final product was 5 hours, with mass loss of 66%.	High potential to convert into fertilisers, animal feed, and feedstock for nutrient recovery processes. However, the product quality was not assessed.	Badaoui et al. (2019)

(Continued)

TABLE 3.5 (Continued)

Examples of Solar Drying Method Used for Drying Various Types of Food Waste

Food Waste and Category	Drying Conditions	Key Findings	Potential Venture	References
Tomato waste	Mixed-mode solar tunnel dryer or open solar drying, 42°C–67°C	Drying time for 5 mm thick tomato waste to achieve equilibrium moisture content: 7 hours by solar tunnel dryer and 15 hours in open solar drying. Drying tomato waste followed Midilli–Kucuk model.	High potential to be converted to fertilisers, animal feed, and feedstock for nutrient recovery processes. The tempering period was ignored in mathematical modelling.	Murugavelh et al. (2019)
Mango waste	Uncontrolled solar drying, 26°C–52°C, relative humidity 42%–61%	Air temperature on the moisture equilibrium influenced the drying of mango seed. A new model was developed to represent uncontrolled solar drying.	High potential to be converted to food ingredients and value-added products such as mango kernel oil, mango kernel butter, mango kernel flour, and biofilms. However, the product quality was not assessed.	Wilkins et al. (2018)
Category C: Others				
Sardine processing waste	Natural or forced convection SD, temperature: 60°C–90°C, air flow 150 or 300 m³/h	In forced convection solar drying, energy consumption decreased with increasing temperature and increased with increasing air flow. The drying curves of sardine heads and their drying time can be represented by Midilli–Kucuk model.	High potential to be converted to feedstock for valorisation to yield flour and oil. However, the product quality was not assessed. In addition, the tempering period was ignored in mathematical modelling.	Bahammou et al. (2019)

employing the active mode (forced convection) solar drying decreased the drying time.

Several relationships between solar drying parameters with drying behaviour have been established in previous research. Based on Wilkins et al. (2018), air temperature on the moisture equilibrium influenced the drying of mango seed. Montero et al. (2011) found that solar drying time decreased with increasing temperature and mass flow. In addition, employing forced convection mode during solar drying will decrease the drying time.

Drying time also depends on the weather condition and types of solar dryer used. For example, open solar drying of *Sargassum* (golden tides) requires 2–3 days in sunny weather and took 7 days in the rainy season to reach the equilibrium moisture content. Drying time for tomato waste to afford a final product at equilibrium moisture content was 5 hours by greenhouse solar dryer (Badaoui et al., 2019), 7 hours by solar tunnel dryer, and 15 hours in open solar drying (Murugavelh et al., 2019).

Several studies concluded that solar drying behaviour for thin-layer food waste can be represented by Midilli and Kucuk (2003). Santos et al. (2014) studied the solar drying of annatto grains and annatto waste grains flour and concluded that Midilli–Kucuk model best represented the drying behaviour in a solar dryer. Grain annatto took longer time to reach moisture content ~5% compared with waste grains flour, with effective diffusivity 6.47×10^{-11} m^2/s for annatto grain and 0.69×10^{-11} for waste grains flour. A similar model was used to describe the behaviour of tomato waste drying via mixed-mode solar tunnel dryer and open solar drying at 42°C–67°C, and sardine processing waste drying via natural and forced convection solar drying at 60°C–90°C. During sun drying, tempering period exists when there is no sun. It was ignored by many researchers. The Midlilli–Kacu model also did not consider the tempering period.

On the other hand, several studies developed their own new model to explain drying of food waste. For example, in a study on the modelling of an uncontrolled solar drying of mango waste in terms of drying rate (Wilkins et al., 2018), equilibrium moisture content should not be ignored in model development despite its notable variations during open solar drying. They concluded that the initial moisture content measurement is critical in drying rate prediction. In addition, effective diffusivity values during the tempering period should be discussed and included in the drying kinetics profile. Further to this, the drying kinetics is not studied together with the product quality. For example, the olive pomace and wastewater have the potential to be converted to valuable products. The quality of dried olive pomace, wastewater, or sludge is not assessed. According to Khdair and Abu-Rumman (2020), the pomace and olive mill wastewater have potential economic values to be converted to fertilisers, valuable anti-oxidants agents, and fatty acids needed in human diet.

3.7 Bio-drying

Bio-drying is a long drying and stabilisation drying method that are famous in biodegradable waste drying to reduce its overall weight and maintaining the calorific value of the processed food waste by relying on microbial to produce heat. Some drying examples are shown in Table 3.6. This is a very useful technique to constrain organics degradation and preserving energy for subsequent utilisation. During these composting processes, bacteria, actinomycetes, fungi, moulds, and yeast oxidise long- and short-chain fatty acids, paper products, and other pollutants served as resources to produce heat. Ventilation strategy, powerful aeration capacity, nutritional properties, and moisture distribution are extremely important to enhance moisture removal efficiently during bio-drying. In addition, a proper selection of ventilation period can control heat, and humidity during the process can improve the microbial degradation and evaporation with adequate effective microbes (Wu et al., 2018). During the bio-drying, the liquid contained in the food waste matrix will be converted to water vapour due to the metabolic heat generated by an aerobic fermentation of organic matter in the food waste. High-moisture organic wastes drying duration is as long as 7–15 days, and the final moisture content ranged from 36% to 65%. According to Tun and Juchelková (2018), this method is good in volume reduction of food wastes, minimised landfill use, and reduced environmental impacts. However, it will produce aqueous and fine particles that cause secondary pollutants. Thus far, there are no effective measures to control the secondary pollution for the fine waste particles. In addition, many more parameters in the bio-drying process need to be studied further with the characteristics of municipal solid waste in different countries. It is important to know how solid waste content influences calorific value during bio-drying and the effects of volatile solids to comprehend the degradation during the process.

Yang and Jahng (2015) dried combined food waste using bio-drying. The food waste was a combination of vermicelli, rice, bean curd, salted radish, salted pepper, carrot, kimchi, fish, and meat with moisture content ranged from 80% to 96%. Response surface methodology (RSM) analysis was adopted to find the optimal operating condition; it was found that the water evaporation was 123.1% and VS degradation was more than 100% under optimal conditions.

Zhao et al. (2010) used bio-drying to dry a mixture of sludge, straw, and sawdust. This mixed bio-drying method significantly improved the drying efficiency and reduced the energy consumption compared to sludge drying alone.

Shao et al. (2010) investigated the effects of bio-drying and waste particle size on heating values, acid gas, and heavy metal emission potential. It was found that the water content of MSW decreased from 73.0% to 48.3%

TABLE 3.6

Examples of Bio-drying Method Used for Drying Various Types of Food Waste

Food Waste and Category	Drying Conditions	Key Findings	Potential Venture	References
Category C: Others				
Municipal solid waste	Ventilated for 5, 10, 15, 20 minutes every 3 hours. Pressure: 3 bar Duration: 14 days	Optimum ventilation: 10 minutes with 82% reduction in moisture content. Half of the electricity cost compared to 20 and 30 minutes.	The solid waste content influences calorific value during bio-drying. The effects of volatile solids to comprehend the degradation during the process. Both are not addressed in the study and have an impact on the product quality.	Ab Jalil et al. (2016)
Mixed food waste MC: 92% Volatile solids: 99% Bulk density (kg/m^3): 464 CV (MJ/kg): 0.11 combined with plastics, bulking agents and paper	Air flow rate: 15 m^3/h Drying duration: 7 days	The moisture content decreased by 80% and calorific values increased by 74% during when combining different waste composition. The optimal conditions were obtained using the Taguchi approach.	The solid waste content influences calorific value during bio-drying. The effects of volatile solids to comprehend the degradation during the process. Both are not addressed in the study and have an impact on product quality.	Mohammed et al. (2019)
Gardening waste (grass, pruning waste, and wood shavings)	Air flow rate: 2–10 L/min Temperature: 22°C–24°C Relative humidity: 55%–75% Duration 20 days	Moisture content reduced to 50%–69% when mixed with bulking agents. The higher flow rate, the higher moisture loss, but the relationship is not linear.	Gardening waste has the potential to be combined with food waste to reduce the moisture content and increase the calorific values and shorten the bio-drying duration.	Colomer Mendoza et al. (2016)
Mixture of sludge, straw and sawdust.		This mixed bio-drying method significantly improved the drying efficiency and reduced the energy consumption compared to sludge drying alone.	Gardening waste has the potential to be combined with this mixture to reduce the moisture content and increase the calorific values and shorten the bio-drying duration.	Zhao et al. (2010)

after bio-drying, whereas its lower heating value (LHV) increased by 157%. In addition, the heavy metal concentrations increased by around 60% due to the loss of dry materials mainly resulting from biodegradation of food residues.

Mohammed et al. (2019) combined the food waste with other wastes during bio-drying and optimised it using the Taguchi method. It was found that the optimum mixtures were food waste: 15 kg; paper: 8 kg; plastic: 10 kg; and bulking agent: 6 kg and food waste: 15 kg; paper: 2 kg; plastic: 10 kg. The food waste significantly reduced the moisture content (MC), whereas Pl was positively correlated to higher calorific values (CV) production.

Colomer Mendoza et al. (2016) dried garden waste using the bio-drying method. It was found that the bulking agents presented the best reduction of moisture after 20 days; the water content of waste was reduced between 50% and 69%, allowing the production of bio-drying food waste with a low heat value (LHV) between 11,063 and 13,709 kJ/kg. However, they are yet to consider to combining with food waste to reduce the moisture content in a shortened period of time.

For bio-drying, mixed drying using straw and sawdust is highly recommended, and the top cover of the bio-drying reactor is recommended to cover with a sponge to avoid heat loss and vapour condensation.

3.8 Others

3.8.1 Supercritical CO_2 Drying of Food Waste

This is an expensive drying method capable of producing brittle and porous materials. A supercritical CO_2 drying is suitable for high-end food waste with high commercial values and highly sensitive to heat. Thus far, no food waste was dried using this method. The supercritical CO_2 (SCO_2) was used to extract high-value bioactive compounds from starch-rich, fibre-rich, beans and vegetables (Yu et al., 2019).

3.8.2 Electrokinetics Drying

The water of food waste can be removed using a constant voltage across a pair of electrodes placed across a volume of solid/liquid slurry. It enhanced the water removal rate. This technique will affect the electromigration, electroosmosis, changes in pH, and electrophoresis. At constant voltage and pressure, the moisture removal rate increased with the magnitude of the voltage applied across the sample. Samples with higher conductivity show higher dewatering efficiency, and the power consumption is also lower than conventional drying methods (Ng et al., 2011).

3.9 Potential Venture and Benefits of Using Smart Drying Method

3.9.1 Smart Drying

Smart drying includes biomimetic systems, artificial intelligence, computer vision technology, microwave dielectric spectroscopy technology, near-infrared reflectance (NIR) spectroscopy, magnetic resonance imaging (MRI), hyper/multi-spectral imaging, ultrasonic imaging, electrostatic sensor technology, and control system for drying environment (Su et al., 2015). A smart dryer can reduce quality degradation and energy wastage. For example, the performance of a dryer is significantly influenced by an automated temperature control system where it is linked to the moisture content of food waste in a smart dryer. The smart drying has a humidity control system that increases the drying speed when the efficiency dropped. Among the smart drying technologies, biomimetic systems and computer vision technology are not suitable for drying of industry food waste as it is focused on improving the product quality such as senses, taste, and appearance for consumable materials (Yousaf et al., 2021). These technologies can be implemented for drying of food waste if it is linked to the temperature, pressure, permittivity of materials, water vapour, humidity, and microbes. Further to this, there is a high requirement for measurement conditions when it involved high operating costs due to large amounts of data and image processing. Below is an example of a possible combination of nature inspiring drying technology using biomimetic design.

3.9.2 Dryer Combined with Biomimetic Design

In microwave dielectric spectroscopy technology, it measures the permittivity of the material as a function of frequency. It can be divided into two distinct methods viz. broadband and resonant systems that operate over a limited frequency range. Broadband systems rely on measurement apparatus operating below its intrinsic electrical resonant frequency. It will not generate far-field radiation. In terms of resonant apparatus, it generates far-field radiation that can be used for transmission measurements (Blakey and Morales-Partera, 2016). To apply this technology in a dryer, a vector network analyser (VNA), an open-ended coaxial cable, or a probe is that measure the electric permittivity of a material stable, a non-absorbent material required. For food waste drying, a frequency of 10^6–10^{14} Hz is suitable for homogeneous food waste material and free space technique with a frequency of 10^9–10^{11} is suitable for inhomogeneous food waste material (el Khaled et al., 2016).

3.9.3 Dryer Combined with Near-Infrared Spectroscopy

Food waste has a wide range of physical characteristics and biochemical compositions. Mallet et al. (2021) design a drying system with a near-infrared spectroscopy to dry 89 subtracts. The authors investigated the complexity of water effects in near-infrared spectroscopy and highlighted the close dependency on the biochemical and physical characteristics of farm waste such as manure, silage, soya meal, and grass using a customised drying system with a close tube loop that consists of a peristaltic pump, strong desiccant like sodium hydroxide at a humidity of 8% at 25°C.

3.9.4 Dryer Combined with Machine Learning

Industry food waste is structurally heterogeneous multi-scale cellular materials. Machine learning-based prediction has the ability to estimate microscale properties including micro-structural changes continuous moisture loss during drying, retention of different bioactive compounds, texture, colour, and any product quality parameters for the dried product. However, complexity is not an issue for machine learning. Most importantly, there are sufficient data to minimise errors. In machine learning-based modelling, information such as micro-scale diffusion, periodic cell rupture, targeted bioactive compounds, and selected texture and colours are required. The most important are machine learning methods such as multi-layer feedforward neural network (MFNN), feedforward neural networks (FNNs), static and recurrent artificial neural networks (ANNs), feedforward neural networks (FNNs), supervisory control and data acquisition (SCADA) systems, ANN and adaptive neuro-fuzzy inference system (ANFIS), and recurrent self-evolving fuzzy neural network (RSEFNN) are used in hot air drying (HAD), solar drying (SD), freeze drying (FD), fluidised bed drying (FBD), rotary drum dryer (RBD), low-pressure super-heated steam drying (LPSSD), MVD, combined HAD-infrared drying with ultrasound and MD of pistachio nuts, seedy grapes, apple, carrot, grain, sugar beet pulp, paneer, tomato, blackberry, red maple, and lignite (Khan et al., 2022). However, machine learning is yet to apply in any industry of food waste drying. Three is an opportunity to build a model that can map a small set of physical variables into a scanning electron microscope image, selected bioactive compounds, colours, or selected textures obtained from experiments to incorporate additional elements like a graphical user interface, etc. Then, let the system allow users to train a specific model using two to four keywords or criteria from human, colour, activity, or images. The results are then keyed in by users into the system for analysis to predict desired parameters. The system would allow users to define data sets, configure training procedures, train the model, test the model, and finally, deploy the model for food waste.

3.9.5 Rotary Dryer Combined with Smart Technology

Perazzini et al. (2014) predicted the residence time distribution (RTD) of solid wastes using a rotary dryer. The food waste used was citrus juice industry organic solid waste. The objectives were to minimise the microbial inactivation, increase the storage time so that it can be applied with recycling, and reuse methods due to it highly perishable condition. Food waste with different characteristics such as sizes, shapes, and densities adopted semi-pilot-scale rotary dryer with ANN method to predict RTD of these homogenous materials and found that the fitting and results are good to consider for a semi-pilot scale. In this study, the number of samples tested was not mentioned, and at conclusion, it was stated that the large and comprehensive database is yet to design at this stage. The size of the database is the main bottleneck of the neuronal method. In addition, Perazzini et al. (2013) used the ANN to model the drying kinetics data. The ANN network developed considered the disturbances such as temperature, velocity, and time to predict the behaviour of the moisture content. This is a small-scale research where 2.0 cm-thick fixed bed filled with moist solid waste was investigated.

3.10 Conclusions

Figure 3.4 shows recommended drying methods for different types of food waste. The decision of selecting suitable method for food waste is based on the type of food waste and product quality. If the dried samples are dedicated for compost or fuel, then the bioactive compounds analysis can be

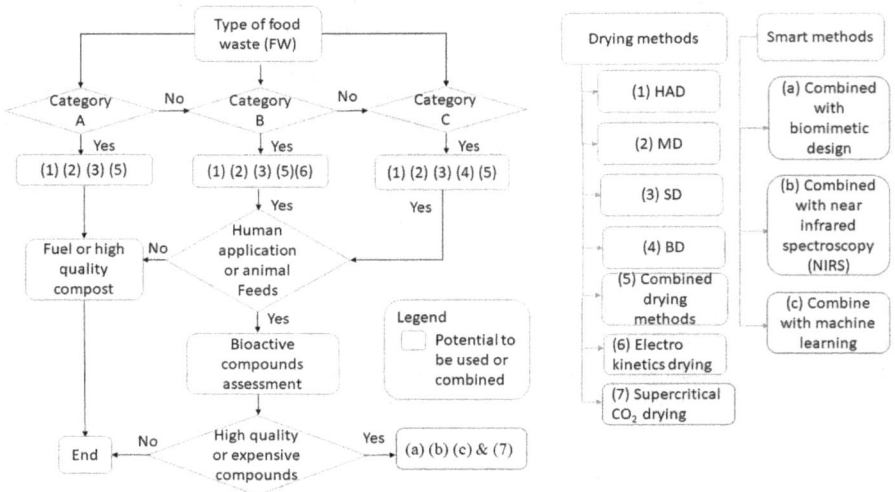

FIGURE 3.4
Recommended drying methods and criteria to consider for different types of food waste. HAD, hot-air drying; MD, microwave drying; SD, sun drying; BD, bio-drying.

ignored and vice versa. In addition, much research did not consider the energy consumptions for methods proposed which has an impact on the economic potential. Further to this, size of food waste materials, feed rate, odour and leachate, possibility of food waste recovery, and dried food waste storage are yet to be explored and considered in food waste drying. In addition, food waste with nutritional values or commercial values can be integrated with smart techniques during drying to predict the product quality of food waste.

References

Ab Jalil, N. A., Basri, H., Basri, N. E. A., & Abushammala, M. F. M. (2016). Biodrying of municipal solid waste under different ventilation periods. *Environmental Engineering Research*, 21(2), 145–151. https://doi.org/10.4491/eer.2015.122

Ahmad, S., Anuar, M. S., Taip, F. S., Shamsudin, R., & Siti Roha, A. M. (2017). Effective moisture diffusivity and activation energy of rambutan seed under different drying methods to promote storage stability. *IOP Conference Series: Materials Science and Engineering*, 203(1). https://doi.org/10.1088/1757-899X/203/1/012025

Ahmad-Qasem, M. H., Cánovas, J., Barrajón-Catalán, E., Carreres, J. E., Micol, V., & García-Pérez, J. v. (2014). Influence of olive leaf processing on the bioaccessibility of bioactive polyphenols. *Journal of Agricultural and Food Chemistry*, 62(26), 6190–6198. https://doi.org/10.1021/jf501414h

Anuar, M. S., Tahir, S. M., Najeeb, M. I., & Ahmad, S. (2019). Banana (*Musa acuminata*) peel drying and powder characteristics obtained through shade and microwave drying processes. *Advances in Materials and Processing Technologies*, 5(2), 181–190. https://doi.org/10.1080/2374068X.2018.1545201

Badaoui, O., Hanini, S., Djebli, A., Haddad, B., & Benhamou, A. (2019). Experimental and modelling study of tomato pomace waste drying in a new solar greenhouse: Evaluation of new drying models. *Renewable Energy*, 133, 144–155. https://doi.org/10.1016/j.renene.2018.10.020

Bahammou, Y., Lamsyehe, H., Kouhila, M., & Lamharrar, A. (2019). Valorization of co-products of sardine waste by physical treatment under natural and forced convection solar drying. *Renewable Energy*, 142, 110–122. https://doi.org/10.1016/j.renene.2019.04.012

Bas-Bellver, C., Barrera, C., Betoret, N., & Seguí, L. (2020). Turning agri-food cooperative vegetable residues into functional powdered ingredients for the food industry. *Sustainability*, 12(4), 1284. https://doi.org/10.3390/su12041284

Blakey, R. T., & Morales-Partera, A. M. (2016). Microwave dielectric spectroscopy – a versatile methodology for online, non-destructive food analysis, monitoring and process control. *Engineering in Agriculture, Environment and Food*, 9(3), 264–273. https://doi.org/10.1016/j.eaef.2016.02.001

Bozkir, H., Tekgül, Y., & Erten, E. S. (2021). Effects of tray drying, vacuum infrared drying, and vacuum microwave drying techniques on quality characteristics and aroma profile of orange peels. *Journal of Food Process Engineering*, 44(1). https://doi.org/10.1111/jfpe.13611

Calín-Sánchez, Á., Lipan, L., Cano-Lamadrid, M., Kharaghani, A., Masztalerz, K., Carbonell-Barrachina, Á. A., & Figiel, A. (2020). Comparison of traditional and novel drying techniques and its effect on quality of fruits, vegetables and aromatic herbs. *Foods, 9*(9), 1261. https://doi.org/10.3390/foods9091261

Chew, Y. M., & King, V. A.-E. (2019). Microwave drying of pitaya (*Hylocereus*) peel and the effects compared with hot-air and freeze-drying. *Transactions of the ASABE, 62*(4), 919–928. https://doi.org/10.13031/trans.13193

Choi, Y., Jung, B., Sung, N., & Han, Y. (2015). A study on the drying characteristics from mixture of food waste and sawdust by using microwave/inner-cycle thermal-air drying process. *Journal of Material Cycles and Waste Management, 17*(2), 359–368. https://doi.org/10.1007/s10163-014-0248-8

Ciccoritti, R., Taddei, F., Gazza, L., & Nocente, F. (2021). Influence of kernel thermal pre-treatments on 5-n-alkylresorcinols, polyphenols and antioxidant activity of durum and einkorn wheat. *European Food Research and Technology, 247*(2), 353–362. https://doi.org/10.1007/s00217-020-03627-4

Colomer Mendoza, F. J., Robles Martínez, F., Piña Guzmán, A. B., Vicente Monserrat, P., & Gallardo Izquierdo, A. (2016). Influence of different airflows and the presence of bulking agent on biodrying of gardeming wastes in reactors. *Revista Internacional de Contaminación Ambiental, 32*(Residuos sólidos), 161–171. https://doi.org/10.20937/RICA.2016.32.05.12

Duarte, Y., Chaux, A., Lopez, N., Largo, E., Ramírez, C., Nuñez, H., Simpson, R., & Vega, O. (2017). Effects of blanching and hot air drying conditions on the physicochemical and technological properties of yellow passion fruit (*Passiflora edulis* var. flavicarpa) by-products. *Journal of Food Process Engineering, 40*(3), e12425. https://doi.org/10.1111/jfpe.12425

el Khaled, D., Castellano, N., Gázquez, J., Perea-Moreno, A.-J., & Manzano-Agugliaro, F. (2016). Dielectric spectroscopy in biomaterials: Agrophysics. *Materials, 9*(5), 310. https://doi.org/10.3390/ma9050310

Fernandez, A., Román, C., Mazza, G., & Rodriguez, R. (2018). Determination of effective moisture diffusivity and thermodynamic properties variation of regional wastes under different atmospheres. *Case Studies in Thermal Engineering, 12*(March), 248–257. https://doi.org/10.1016/j.csite.2018.04.015

França, F., Harada-Padermo, S. d. S., Frasceto, R. A., Saldaña, E., Lorenzo, J. M., Vieira, T. M. F. de S., & Selani, M. M. (2022). Umami ingredient from shiitake (Lentinula edodes) by-products as a flavor enhancer in low-salt beef burgers: Effects on physicochemical and technological properties. *LWT, 154*, 112724. https://doi.org/10.1016/j.lwt.2021.112724

Fu, B. A., & Chen, M. Q. (2020). Microwave drying performance of spent coffee grounds briquette coupled with mineral additives. *Drying Technology, 38*(15), 2094–2101. https://doi.org/10.1080/07373937.2019.1692862

Ghasemi, A., Chayjan, R. A., & Najafabadi, H. J. (2018). Optimization of granular waste production based on mechanical properties. *Waste Management, 75*, 82–93. https://doi.org/10.1016/j.wasman.2018.02.019

Gómez-de la Cruz, F. J., Palomar-Carnicero, J. M., Hernández-Escobedo, Q., & Cruz-Peragón, F. (2020). Determination of the drying rate and effective diffusivity coefficients during convective drying of two-phase olive mill waste at rotary dryers drying conditions for their application. *Renewable Energy, 153*, 900–910. https://doi.org/10.1016/j.renene.2020.02.062

González, C. M., Hernando, I., & Moraga, G. (2021). Influence of ripening stage and de-astringency treatment on the production of dehydrated persimmon snacks. *Journal of the Science of Food and Agriculture, 101*(2), 603–612. https://doi.org/10.1002/jsfa.10672

Hirunrattana, P., & Limpisophon, K. (2019). Production of calcium-rich snack from salmon bone. *Italian Journal of Food Science, SI*, 192–197.

Höglund, E., Eliasson, L., Oliveira, G., Almli, V. L., Sozer, N., & Alminger, M. (2018). Effect of drying and extrusion processing on physical and nutritional characteristics of bilberry press cake extrudates. *LWT, 92*, 422–428. https://doi.org/10.1016/j.lwt.2018.02.042

Khaloahmadi, A., Borghei, A. M., & roustpoor, O. R. (2021). Evaluate the drying of food waste using cabinet dryer. 13 September 2021, PREPRINT (Version 1) available at Research Square: https://doi.org/10.21203/rs.3.rs-874515/v1

Khan, M. I. H., Sablani, S. S., Joardder, M. U. H., & Karim, M. A. (2022). Application of machine learning-based approach in food drying: Opportunities and challenges. *Drying Technology, 40*(6), 1051–1067. https://doi.org/10.1080/07373937.2020.1853152

Khdair, A., & Abu-Rumman, G. (2020). Sustainable environmental management and valorization options for olive mill byproducts in the Middle East and North Africa (MENA) region. *Processes, 8*(6), 671. https://doi.org/10.3390/pr8060671

Kieu Tran, T. M., Kirkman, T., Nguyen, M., & van Vuong, Q. (2020). Effects of drying on physical properties, phenolic compounds and antioxidant capacity of Robusta wet coffee pulp (Coffea canephora). *Heliyon, 6*(7), e04498. https://doi.org/10.1016/j.heliyon.2020.e04498

Ko, H. Y., Shin, D. H., Oh, J. H., Kang, H., Choi, D. E., & Choi, S. (2021). Numerical simulation of thermal flow characteristics in plasma reactor for rotten citrus fruits drying. *Applied Science and Convergence Technology, 30*(3), 70–73. https://doi.org/10.5757/ASCT.2021.30.3.70

Kobayashi, N., Hamabe, H., Yamaji, S., Suami, A., & Itaya, Y. (2019). Effect of sludge char addition on drying rate and decomposition rate of organic waste during bio-drying. *Journal of Material Cycles and Waste Management, 21*(4), 897–904. https://doi.org/10.1007/s10163-019-00847-z

Koukouch, A., Idlimam, A., Asbik, M., Sarh, B., Izrar, B., Bostyn, S., Bah, A., Ansari, O., Zegaoui, O., & Amine, A. (2017). Experimental determination of the effective moisture diffusivity and activation energy during convective solar drying of olive pomace waste. *Renewable Energy, 101*, 565–574. https://doi.org/10.1016/j.renene.2016.09.006

Kumar, N., Sarkar, B. C., & Sharma, H. K. (2011). Effect of air velocity on kinetics of thin layer carrot pomace drying. *Food Science and Technology International, 17*(5), 459–469. https://doi.org/10.1177/1082013211398832

Lin, X., Xu, J.-L., & Sun, D.-W. (2021). Comparison of moisture uniformity between microwave-vacuum and hot-air dried ginger slices using hyperspectral information combined with semivariogram. *Drying Technology, 39*(8), 1044–1058. https://doi.org/10.1080/07373937.2020.1741006

Liu, H., E, J., Ma, X., & Xie, C. (2016). Influence of microwave drying on the combustion characteristics of food waste. *Drying Technology, 34*(12), 1397–1405. https://doi.org/10.1080/07373937.2015.1118121

Ma, J., Zhang, L., Mu, L., Zhu, K., & Li, A. (2018). Thermally assisted bio-drying of food waste: Synergistic enhancement and energetic evaluation. *Waste Management*, *80*, 327–338. https://doi.org/10.1016/j.wasman.2018.09.023

Mallet, A., Charnier, C., Latrille, É., Bendoula, R., Steyer, J.-P., & Roger, J.-M. (2021). Unveiling non-linear water effects in near infrared spectroscopy: A study on organic wastes during drying using chemometrics. *Waste Management*, *122*, 36–48. https://doi.org/10.1016/j.wasman.2020.12.019

Manzoor, S., Yusof, Y. A., Ling, C. N., Syafinaz, I., Amin Tawakkal, M., Fikry, M., & Sin, C. L. (2019). Science & technology thin-layer drying characteristics of papaya (Carica papaya) peel using convection oven and microwave drying. *Pertanika Journal of Science & Technology*, *27*(3), 1207–1226

Midilli, A., & Kucuk, H. (2003). Mathematical modeling of thin layer drying of pistachio by using solar energy. *Energy Conversion and Management*, *44*(7), 1111–1122. https://doi.org/10.1016/S0196-8904(02)00099-7

Milanovic, M., Komatina, M., Zlatanovic, I., Manic, N., & Antonijevic, D. (2021). Kinetic parameters identification of conductive enhanced hot air drying process of food waste. *Thermal Science*, *25*(3 Part A), 1795–1807. https://doi.org/10.2298/TSCI200312223M.

Milledge, J. J., & Harvey, P. J. (2016). Golden tides: Problem or golden opportunity? The valorisation of Sargassum from beach inundations. *Journal of Marine Science and Engineering*, *4*(3). https://doi.org/10.3390/jmse4030060

Mohammed, M., Ozbay, I., Karademir, A., & Donkor, A. (2019). Effect of waste matrix for the optimization of moisture content and calorific value of biodried material using Taguchi DOE. *Drying Technology*, *37*(11), 1352–1362. https://doi.org/10.1080/07373937.2018.1500484

Mohan, A., Shanmugam, S., & Nithyalakshmi, V. (2016). Comparison of the nutritional, physico-chemical and anti-nutrient properties of freeze and hot air dried watermelon (Citrullus lanatus) rind. *Biosciences, Biotechnology Research Asia*, *13*(2), 1113–1119. https://doi.org/10.13005/bbra/2140

Montero, I., Miranda, T., Arranz, J. I., & Rojas, C. V. (2011). Thin layer drying kinetics of by-products from olive oil processing. *International Journal of Molecular Sciences*, *12*(11), 7885–7897. https://doi.org/10.3390/ijms12117885

Moussaoui, H., Bahammou, Y., Tagnamas, Z., Kouhila, M., Lamharrar, A., & Idlimam, A. (2021). Application of solar drying on the apple peels using an indirect hybrid solar-electrical forced convection dryer. *Renewable Energy*, *168*, 131–140. https://doi.org/10.1016/j.renene.2020.12.046

Murugavelh, S., Anand, B., Midhun Prasad, K., Nagarajan, R., & Azariah Pravin Kumar, S. (2019). Exergy analysis and kinetic study of tomato waste drying in a mixed mode solar tunnel dryer. *Energy Sources, Part A: Recovery, Utilization, and Environmental Effects*, 1–17. https://doi.org/10.1080/15567036.2019.1679289

Mutlu, Ö. Ç., Büchner, D., Theurich, S., & Zeng, T. (2021). Combined use of solar and biomass energy for sustainable and cost-effective low-temperature drying of food processing residues on industrial-scale. *Energies*, *14*(3), 561. https://doi.org/10.3390/en14030561

Nagel, A., Neidhart, S., Anders, T., Elstner, P., Korhummel, S., Sulzer, T., Wulfkühler, S., Winkler, C., Qadri, S., Rentschler, C., Pholpipattanapong, N., Wuthisomboon, J., Endress, H.-U., Sruamsiri, P., & Carle, R. (2014). Improved processes for the conversion of mango peel into storable starting material for the recovery of functional co-products. *Industrial Crops and Products*, *61*, 92–105. https://doi. org/10.1016/j.indcrop.2014.06.034

Nagel, A., Neidhart, S., Kuebler (née Wulfkuehler), S., Elstner, P., Anders, T., Korhummel, S., Sulzer, T., Kienzle, S., Winkler, C., Qadri, S., Rentschler, C., Pholpipattanapong, N., Wuthisomboon, J., Endress, H.-U., Sruamsiri, P., & Carle, R. (2017). Applicability of fruit blanching and intermittent microwave-convective belt drying to industrial peel waste of different mango cultivars for the recovery of functional coproducts. *Industrial Crops and Products*, *109*, 923–935. https://doi.org/10.1016/j.indcrop.2017.08.028

Namir, M., Elzahar, K., Ramadan, M. F., & Allaf, K. (2017). Cactus pear peel snacks prepared by instant pressure drop texturing: Effect of process variables on bioactive compounds and functional properties. *Journal of Food Measurement and Characterization*, *11*(2), 388–400. https://doi.org/10.1007/s11694-016-9407-z

Ng, S. K., Plunkett, A., Stojceska, V., Ainsworth, P., Lamont-Black, J., Hall, J., White, C., Glendenning, S., & Russell, D. (2011). Electro-kinetic technology as a low-cost method for dewatering food by-product. *Drying Technology*, *29*(14), 1721–1728. https://doi.org/10.1080/07373937.2011.602199

Nieto-Calvache, J. E., de Escalada Pla, M., & Gerschenson, L. N. (2019). Dietary fibre concentrates produced from papaya by-products for agroindustrial waste valorisation. *International Journal of Food Science and Technology*, *54*(4), 1074–1080. https://doi.org/10.1111/ijfs.13962

Noori, A. W., Royen, M. J., Medveďová, A., & Haydary, J. (2022). Drying of food waste for potential use as animal feed. *Sustainability*, *14*(10), 5849. https://doi. org/10.3390/su14105849

Ostermann-Porcel, M. V., Rinaldoni, A. N., Rodriguez-Furlán, L. T., & Campderrós, M. E. (2017). Quality assessment of dried okara as a source of production of gluten-free flour. *Journal of the Science of Food and Agriculture*, *97*(9), 2934–2941. https://doi.org/10.1002/jsfa.8131

Pasten, A., Uribe, E., Stucken, K., Rodríguez, A., & Vega-Gálvez, A. (2019). Influence of drying on the recoverable high-value products from olive (cv. Arbequina) waste cake. *Waste and Biomass Valorization*, *10*(6), 1627–1638. https://doi. org/10.1007/s12649-017-0187-4

Perazzini, H., Freire, F. B., & Freire, J. T. (2013). Drying kinetics prediction of solid waste using semi-empirical and artificial neural network models. *Chemical Engineering & Technology*, *36*(7), 1193–1201. https://doi.org/10.1002/ceat.201200593

Perazzini, H., Freire, F. B., & Freire, J. T. (2014). Prediction of residence time distribution of solid wastes in a rotary dryer. *Drying Technology*, *32*(4), 428–436. https:// doi.org/10.1080/07373937.2013.835317

Plazzotta, S., Calligaris, S., & Manzocco, L. (2018). Application of different drying techniques to fresh-cut salad waste to obtain food ingredients rich in antioxidants and with high solvent loading capacity. *LWT*, *89*, 276–283. https://doi. org/10.1016/j.lwt.2017.10.056

Pratap Singh, A., Mandal, R., Shojaei, M., Singh, A., Kowalczewski, P. Ł., Ligaj, M., Pawlicz, J., & Jarzębski, M. (2020). Novel drying methods for sustainable upcycling of brewers' spent grains as a plant protein source. *Sustainability*, *12*(9), 3660. https://doi.org/10.3390/su12093660

Ran, X., Zhang, M., Wang, Y., & Adhikari, B. (2019). Novel technologies applied for recovery and value addition of high value compounds from plant byproducts:

A review. *Critical Reviews in Food Science and Nutrition, 59*(3), 450–461. https://doi.org/10.1080/10408398.2017.1377149

Rodriguez, A., Sancho, A. M., Barrio, Y., Rosito, P., & Gozzi, M. S. (2019). Combined drying of nopal pads (*Opuntia ficus-indica*): Optimization of osmotic dehydration as a pretreatment before hot air drying. *Journal of Food Processing and Preservation, 43*(11). https://doi.org/10.1111/jfpp.14183

Şahin, S., Elhussein, E., Bilgin, M., Lorenzo, J. M., Barba, F. J., & Roohinejad, S. (2018). Effect of drying method on oleuropein, total phenolic content, flavonoid content, and antioxidant activity of olive (*Olea europaea*) leaf. *Journal of Food Processing and Preservation, 42*(5), e13604. https://doi.org/10.1111/jfpp.13604

Salina MD Salim, N., Kidakasseril Kurian, J., Gariepy, Y., & Raghavan, V. (2016). Application and the techno-economical aspects of integrated microwave drying systems for development of dehydrated food products. *Japan Journal of Food Engineering, 17*(4), 139–146. https://doi.org/10.11301/jsfe.17.139

Santos, D. d. C., Queiroz, l. J. d. M., Figueirêdo, R. M. F. d., & Oliveira, E. N. A. d. (2014). Solar drying of annatto grains and waste grains flour of annatto. *Bioscience Journal, 30*(2), 436–446.

Scalcon, A., Sone, A. P., Johann, G., Gimenes, M. L., & Vieira, M. G. A. (2018). Shrinkage of digested sludge from gelatin production. *Drying Technology, 36*(13), 1603–1618. https://doi.org/10.1080/07373937.2017.1419479

Serrano-Díaz, J., Sánchez, A. M., Alvarruiz, A., & Alonso, G. L. (2013). Preservation of saffron floral bio-residues by hot air convection. *Food Chemistry, 141*(2), 1536–1543. https://doi.org/10.1016/j.foodchem.2013.04.029

Shao, L.-M., Ma, Z.-H., Zhang, H., Zhang, D.-Q., & He, P.-J. (2010). Bio-drying and size sorting of municipal solid waste with high water content for improving energy recovery. *Waste Management, 30*(7), 1165–1170. https://doi.org/10.1016/j.wasman.2010.01.011

Shih, Y., Wang, W., Hasenbeck, A., Stone, D., & Zhao, Y. (2020). Investigation of physicochemical, nutritional, and sensory qualities of muffins incorporated with dried brewer's spent grain flours as a source of dietary fiber and protein. *Journal of Food Science, 85*(11), 3943–3953. https://doi.org/10.1111/1750-3841.15483

Sinrod, A. J. G., Avena-Bustillos, R. J., Olson, D. A., Crawford, L. M., Wang, S. C., & McHugh, T. H. (2019). Phenolics and antioxidant capacity of pitted olive pomace affected by three drying technologies. *Journal of Food Science, 84*(3), 412–420. https://doi.org/10.1111/1750-3841.14447

Sogi, D. S., Siddiq, M., Greiby, I., & Dolan, K. D. (2013). Total phenolics, antioxidant activity, and functional properties of 'Tommy Atkins' mango peel and kernel as affected by drying methods. *Food Chemistry, 141*(3), 2649–2655. https://doi.org/10.1016/j.foodchem.2013.05.053

Sothornvit, R. (2012). Drying process and mangosteen rind powder product. *Acta Horticulturae, 928*, 233–241. https://doi.org/10.17660/ActaHortic.2012.928.29

Sridhar, K., & Charles, A. L. (2020). Mathematical modeling and effect of drying temperature on physicochemical properties of new commercial grape "Kyoho" seeds. *Journal of Food Process Engineering, 43*(3). https://doi.org/10.1111/jfpe.13203

Su, Y., Zhang, M., & Mujumdar, A. S. (2015). Recent developments in smart drying technology. *Drying Technology, 33*(3), 260–276. https://doi.org/10.1080/07373937.2014.985382

Tun, M. M., & Juchelková, D. (2018). Drying methods for municipal solid waste quality improvement in the developed and developing countries: A review. *Environmental Engineering Research*, 24(4), 529–542. https://doi.org/10.4491/eer.2018.327

Tylewicz, U., Nowacka, M., Rybak, K., Drozdzal, K., Dalla Rosa, M., & Mozzon, M. (2020). Design of healthy snack based on kiwifruit. *Molecules*, 25(14), 3309. https://doi.org/10.3390/molecules25143309

Uribe, E., Lemus-Mondaca, R., Vega-Gálvez, A., López, L. A., Pereira, K., López, J., Ah-Hen, K., & di Scala, K. (2013). Quality characterization of waste olive cake during hot air drying: Nutritional aspects and antioxidant activity. *Food and Bioprocess Technology*, 6(5), 1207–1217. https://doi.org/10.1007/s11947-012-0802-0

Valadez-Carmona, L., Plazola-Jacinto, C. P., Hernández-Ortega, M., Hernández-Navarro, M. D., Villarreal, F., Necoechea-Mondragón, H., Ortiz-Moreno, A., & Ceballos-Reyes, G. (2017). Effects of microwaves, hot air and freeze-drying on the phenolic compounds, antioxidant capacity, enzyme activity and microstructure of cacao pod husks (Theobroma cacao L.). *Innovative Food Science & Emerging Technologies*, 41, 378–386. https://doi.org/10.1016/j.ifset.2017.04.012

Wallin, E., Fornell, R., Räftegård, O., Walfridson, T., & Benson, J. (2020). Design and integration of heat recovery in combination with solar and biomass-based heating in a drying plant. *Chemical Engineering Transactions*, 81, 1387–1392. https://doi.org/10.3303/CET2081232

Wilkins, R., Brusey, J., & Gaura, E. (2018). Modelling uncontrolled solar drying of mango waste. *Journal of Food Engineering*, 237, 44–51. https://doi.org/10.1016/j.jfoodeng.2018.05.012

Wu, Z.-Y., Cai, L., Krafft, T., Gao, D., & Wang, L. (2018). Biodrying performance and bacterial community structure under variable and constant aeration regimes during sewage sludge biodrying. *Drying Technology*, 36(1), 84–92. https://doi.org/10.1080/07373937.2017.1301951

Yang, B., & Jahng, D. (2015). Optimization of food waste bioevaporation process using response surface methodology. *Drying Technology*, 33(10), 1188–1198. https://doi.org/10.1080/07373937.2014.943235

Yousaf, K., Chen, K., & Azam Khan, M. (2021). An introduction of biomimetic system and heat pump technology in food drying industry. In Maki K. Habib and César Martín-Gómez (Eds.), *Biomimetics*. IntechOpen. https://doi.org/10.5772/intechopen.93386

Yu, I. K. M., Attard, T. M., Chen, S. S., Tsang, D. C. W., Hunt, A. J., Jérôme, F., Ok, Y. S., & Poon, C. S. (2019). Supercritical carbon dioxide extraction of value-added products and thermochemical synthesis of platform chemicals from food waste. *ACS Sustainable Chemistry & Engineering*, 7(2), 2821–2829. https://doi.org/10.1021/acssuschemeng.8b06184

Zhao, L., Gu, W.-M., He, P.-J., & Shao, L.-M. (2010). Effect of air-flow rate and turning frequency on bio-drying of dewatered sludge. *Water Research*, 44(20), 6144–6152. https://doi.org/10.1016/j.watres.2010.07.002

4

Selection of Drying Technologies Based on Waste Characteristics and Valorisation Routes

4.1 Introduction

Drying technology plays a very important role in reducing pathogens and increasing shelf life by reducing water activity to a safe level. It will be lighter and easy to store and transport. High-quality food waste with nutritional values can be used for human nutrition. Low-nutritional-value food waste will be used as animal feed. In addition, it can be used as fertiliser for the plantation industry as well. Therefore, criteria to consider when purchasing or designing a food waste dryer are important.

The selection of industry food waste dryers is shown in Figure 4.1. Confirming the size, density, intended final moisture content, and reaction to heat on the food waste is important. From a business point of view, the supply of raw materials to meet the production requirement and efficiency of the dryer, capital cost, and annualised operating cost is required to be considered in selecting a suitable dryer. Food waste has different sizes, densities, and characteristics. For example, it can be fine powder with non-uniform, small to large particle size that has a different appearance, form, sludge form, small- to large-size skin form, wet form, dry form, or sticky form. Therefore, fine powder, sludge, large-size food wastes required different types of dryers because under-sized dryers will lead to clogging and product not properly dried and quality control issue; meanwhile the oversized will lead to high fuel consumption, over-dried, and quality control issue. The density of food waste is important as well as it can affect the flow of food waste through the dryer as it is aligned with the design of the conveyor belt. Further to this, all food wastes have different characteristics. Some food wastes are sludgy and sticky while wet. It is important to know how to load and unload into the dryer and prevent plugging on the processing line using non-stick surfaces, weirs, and paddles to keep material flowing. After drying, it should be easy to clean by ensuring multiple fittings are used in the dryer design that allows

DOI: 10.1201/9781003312802-4

99

Checklist in selection of industry food waste dryer

Size	Density	Final moisture content of food waste.	Reaction to heat	Production requirement	Efficiency and cost

FIGURE 4.1
The selection criteria of industry food waste dryer.

Proposed location	Requester (Name)	
	Date	
	Department	
	Equipment Details	
	Equipment Name	
	Purpose	
	Equipment dimension	L x W x H (m)
		Weight (Ton)
	Electricity	Yes ☐ No ☐
	Water supply	Yes ☐ No ☐
	Compressed air	Yes ☐ No ☐
	Gas bottle	Yes ☐ No ☐
	Air extraction system	Yes ☐ No ☐
	Proposed Location	
	Room No	
	Area Name	
	Duration	
	Location authorisation	
	Plant Manager (Name): Signature:	
	Process Engineer (Name): Signature:	
	Person In Charge (Name): Signature:	
	Equipment Photo	
	Remarks	Yes ☐ No ☐

FIGURE 4.2
Example of food waste dryer floor plan.

the removal of potential build-up. If the food waste is very coarse, reinforced materials with abrasive resistance are required.

The moisture content of food waste is also very important. A lab-scale analysis is recommended to determine the drying kinetics and characteristics of food waste. The moisture content of food waste can be determined by setting the final moisture content. Different food waste initial moisture content, final moisture content, and equilibrium moisture content for different settings. Drying characteristics and drying kinetics for different settings for different food waste. If required to remove the internal moisture content, different drying methods are required. Further to this, either continuous or batch process, space, power supply (usually 415 V, three-phase), and floor plan (Figure 4.2) need to be confirmed as heavy equipment with vibration. A hard-standing non-porous surface is required to place this dryer. The size of the dryer (length, width, height), the batch of drying per

TABLE 4.1

Economic Evaluation Template for Drying of Food Waste

Item	Value ($)
Annual pay for disposal of food waste	
Capital expenditure for unit operations	
Annual utilities costs	
Annual gross-saving	
Annual net-saving	
Return on investment	Preferably more than 30%
Breakeven point	Preferably less than 3 years

year, and the hour per batch. The fine powder types of food waste need ventilation, louvre, ductwork, and baghouse filtration. Drying food waste involved heating. Some food waste can cause chemical reactions when heated up and it may melt or burn. Therefore, the residence time becomes very important as it can reduce it at high heat. In addition, it is important to decide whether to use direct or indirect heating to minimise the reaction. In converting the food waste to biogas production, high temperature is acceptable as it increases the reaction kinetics that can lead to higher microorganisms being affected by temperature; reaction kinetics brings digestion allowing higher reaction kinetics and higher biogas production (Slopiecka et al., 2022).

Evaluation of dryer efficiency and cost analysis is required. A cash flow for drying of food waste must include the annual pay for food waste disposal, capital expenditure for unit operation, annual utility costs, annual gross-saving, annual net-saving, return on investment (ROI), and breakeven point. This is because the upfront capital expenses, operating costs, life spans, and types of fuel will affect the feasibility of the process. The economic evaluation template is shown in Table 4.1.

4.2 Valorisation of Food Waste: Examples from Different Industries

Table 4.2 summarises efforts in reducing food waste by food-processing industries. Zhenjiang Zhinong Food Co. Ltd. processes the grain dust that cannot be recycled into the product. The food waste was 3% of grain dust from the production process. The company processes the grain dust into animal feed and sold to professional farmers. Valorising the food waste increased the company's income and avoids the waste of resources. The recycling of grain dust prevents dust from entering the atmosphere and causing air dust pollution.

The company adopted microwave drying and hot-air drying to reduce food waste or food by-products. According to the scale of the factory, the capacity production of microwave drying is 1.5 tons/day (Figures 4.3 and 4.4), and the hot-air drying is 300 kg/h. The challenge is that microbes are not easy to control, and the dried product has a poor flavour (beany flavour). Figures 4.5 and 4.6 show operating conditions of the food waste process. According to the company person in charge, relevant specialists must have professional theoretical knowledge and practical experience; on the one hand, the specialist can use their professional knowledge to improve the product yield and

FIGURE 4.3
A microwave dryer used by Zhenjiang Zhinong Food Co. Ltd.

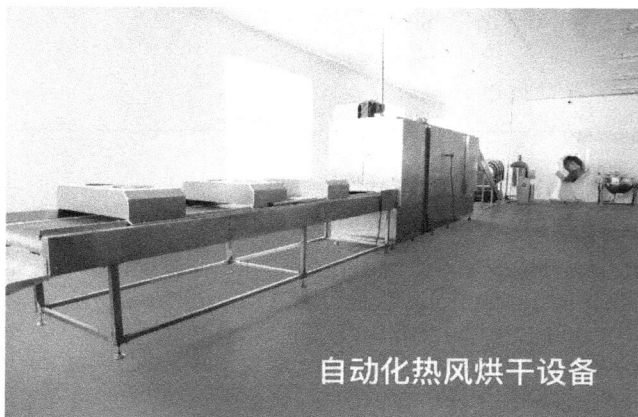

FIGURE 4.4
A hot-air dryer used by Zhenjiang Zhinong Food Co. Ltd.

FIGURE 4.5
Food waste processing photo from Zhenjiang Zhinong Food Co. Ltd.

FIGURE 4.6
Packaging of food waste processing photo from Zhenjiang Zhinong Food Co. Ltd.

reduce loss. Further to this, the product waste is reprocessed with high value. Now many experts are interested in the comprehensive utilisation of food waste and food by-products.

Industry 4.0/5.0 is the era of intelligence and personalised customisation. Intelligent manufacturing mode is realised through the real-time inter-connection of people, equipment, and products. First, according to the order quantity, artificial intelligence directly connected design and production with a highly automated production process, to reduce the waste of resources and the generation of waste to the greatest extent. Secondly, for by-products or food waste generated in the process of production, it can realise the application of industry 4.0/5.0 in the drying and valorisation of food waste or food by-products through personalised product customisation and digital support.

Since the year 2015, Nestlé has taken several actions for all branches to achieve landfill-free status, reducing operational food waste worldwide (Nestlé, 2023). In the year 2015, Nestlé succeeded in contributing zero waste to landfill, while Nestlé Malaysia also reached this goal by the year 2017, through different ways of food waste recycling such as conversion into animal feed and fertiliser which involve solar and thermal drying (Burge, 2015; Nestlé Malaysia, 2020).

Other than the well-known food-processing industries, many entrepreneurs such as NETZRO, Renewal Mill, and Barnana have committed themselves in food upcycling, reducing the generation of food waste. The drying technologies used are shown in Table 4.2. NETZRO and Renewal Mill convert the by-products generated by food-processing industries into value-added products such as nutritional flour, whereas Barnana utilises the "imperfect" bananas, which are typically eliminated from the food-processing chain to produce banana-flavoured snacks (Barnana, 2023; Mill, 2023; NETZRO, 2023).

PepsiCo recycled snack food waste into energy and fertiliser. It generated 35% of the electricity needed to run the Frito-Lay food-processing plants. The major food waste mixtures are potato peels and corn kernels. The company outsources the food waste sludge to an external partner for about 3 weeks for drying purposes. At this moment, only focus on the final processing stage to increase the nutritional values. Kraft Heinz has five new manufacturing facilities that achieved zero-waste-to-landfill status in 2021. The total waste produced in 2021 is about 82 kMT, which is about 1.1% of waste per production. Food wastes are potato peeling and corn husk. It was reported that the Frito-Lays factory in Arizona burns leftover sludge from the water filters technology to produce methane gas (Martin, 2007). It will be supplied for the plant's boiler. It has also been reported that the recovering potato starch was dried using waste heat and then sold to chemical companies to use in their production line (Greenfield, 2009).

In France, a company called Lctyos turned fish skin for restaurant waste into leather. The fish skin was dried using room-temperature air drying. KM Sugar Mill Ltd. owned a bagasse-based co-gen power plant. Bagasse waste generated is burned to produce electrical, and it can be considered a low-pollution alternative to fossil fuel. In India, Yash Pakka, India paper mills convert sugar cane waste into paper, food containers, and utensils. José Alberto Aram Berri from Argentina invented an innovative alternative to firewood and charcoal. He dried waste pomace into logs using air drying under the sun. Alhaji Siraj Bah owned a company called Rugsal Trading in Freetown. The company produced Bio Briquettes with about 100 metric tons of inventory. The drying duration was about a week before sending it to fire in a steam drum.

TABLE 4.2

Efforts in Reducing the Food Waste by Food-Processing Industries

Company Name	Efforts in Reducing the Food Waste	Drying Methods	Achievements	References
Nestlé	Food waste prevention, minimisation, and valorisation. Organic food waste was converted into animal feed and organic fertilisers and converted food waste into biofuel. The company also raised awareness on food waste reduction among the consumers, provided training to employees to encourage leftovers reduction, and created partnerships with business cooperation and green parties such as UN-FAO and European Programme.	Solar drying and thermal drying	95% of the sites have achieved "zero landfill" status. Nestlé UK and Ireland have reduced 41% of operational food waste.	Nestlé (2015, 2023)
PepsiCo	The company minimised the generation of food waste by optimising the efficiency of manufacturing process. In addition, recycled of food waste by converting potato peelings to low-carbon fertiliser and generated biogas via anaerobic digestion. Created partnership with professionals to work on food waste recycling, composting, and waste-to-energy operations.	N/S	98% of waste is recycled. The conversion of food waste into fertiliser not only avoids the disposal cost but also reduces the cost of purchasing chemical fertilisers.	Heather Clancy (2013) and Pepsico (n.d.)
Tyson Food	Conversion of food waste into saleable by-products. Food waste is being converted into animal feed, biofuels, and fertiliser.	N/S	80% of waste is recycled in year 2020	Tyson (2022)
Kraft Heinz Company	Recycling of food waste. Conversion of food waste to animal feed. Minimising generation of food waste. Food donation. Reporting. Consistent monitoring of food waste reduction progress.	Solar drying	10% of the manufacturing facilities has achieved "zero-landfilling" status in year 2019	Kraft Heinz (2022)

(Continued)

TABLE 4.2 (*Continued*)

Efforts in Reducing the Food Waste by Food-Processing Industries

Company Name	Efforts in Reducing the Food Waste	Drying Methods	Achievements	References
Suntory	Recycling of food waste. Food waste is converted into animal feed and fertiliser.	N/S	Achieved 100% recycling rate since year 2014	Suntory (n.d.)
NETZRO	Recycling of eggshells. Separation of membrane and calcium hard shell: • membrane can be utilised in cosmetic products due to the presence of collagen • calcium hard shell can be utilised in production of supplements and antacids due to rich content in calcium carbonate Conversion of food waste into soil amendments Brewery spent grain is converted into flour	Infrared drying	A food upcycling platform to help in reducing food waste	Jossi (2018) and NETZRO (n.d.)
Renewal Mill	Conversion of soybean pulp, oat pulp, and almond pulp into flour	Thermal drying	An upcycled food company fighting global food loss	Eric Peterson (2019) and Renewal Mill (n.d.)

Note: N/S: The drying technology applied is not specified.

In India, banana trunk waste was turned into menstrual pads (McCauley, 2022). They cut the stalk in half, split it layer by layer, and put it in the machine to get the fibre before drying. Extraction units were set up for farmers to generate income by producing fibre at the start of the circular economy. Liquid from the stem will be produced as fertiliser. However, it is very costly to request farmers to invest in extraction units. The same goes for dryers; the same principles are required to apply to let farmers generate income using agricultural waste or else they will just use conventional drying techniques. Currently, the existing large-scale food waste is reprocessed and turned into new products using conventional hot-air drying or sun drying. The duration is extremely long, and it is hard to reduce the pollution generated during the process. A suitable dryer should be purchased and located in the different areas based on the food waste.

The edible or non-edible fruits and vegetables, spices, sugarcane, bakeries, confectioneries, oilseeds, beverages, milk and milk-based products, egg, meat, and seafood can be processed into new products. Food waste can also be obtained from peels, cooking oil, bones, discharged food portions, spoiled/contaminated supplies, skins, and outer coverings including fruit and vegetable trimmings, seeds, and pulp.

Various studies to investigate the recovery of food-processing by-products or food wastes that have low deterioration levels and availability are conducted. Food wastes such as skins, husks, vegetable and fruit peels, seeds, animal meat, bones, or eggshells contain high-value reusable materials. However, the fundamental concept of valorisation of food-processing waste is not well addressed. For example, a guide to the principles of animal feeds and alternative foods production using food waste is not visible, and the detailed concepts of converting food waste to biomaterials and compost are not well documented.

Example 1: Grape Pomace

Grape pomace or marc is the residue after grape pressing. The raw grape marc is commonly direct spreading to land and then ploughed and cropped, as an alternative to the composting system that may potentially cause environmental damage. Several methods to dispose of the grape pomace, for example, direct spreading to land, drying to produce valuable products, composting, combustion to produce electric power supply, gasification to produce heat and electricity power, and pyrolysis to produce biochar and excess heat for heating equipment. Grape pomace is generated from the seed, stalks, and skins after the pressing process. The growing of wine production causes increase in the grape pomace generation. The direct spreading of food wastes to land could not be economically viable and may cause economic and environmental constraints.

The intermediate processes involve phenolic extraction, ensiling to recover secondary alcohol, acid hydrolysis to generate a range of useful ligno-cellulosic

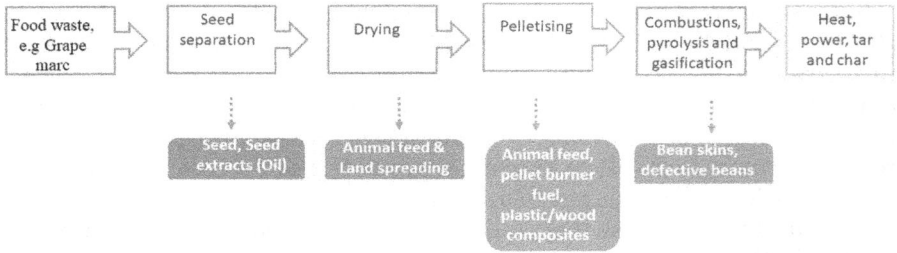

FIGURE 4.7
Repurposing grape marc (Jones, Sarah McLaren, et al., 2020).

compounds or as a precursor enzymatic saccharification, and follow up with the fermentation to produce bioethanol, or biogas production by anaerobic digestion. Figure 4.7 illustrates parts of the drying process of grape pomace to generate valuable products (Jones et al., 2020).

The pressing of the grapes, with an initial moisture content of 67%, is operated at 2 bar in two press cycles during the winery processing stage. An operating pressure higher than 2 bar is not desirable to prevent unpleasant flavour detected in the juice sample. The removal of the moisture content through mechanical dewatering process is more energy efficient as compared with the evaporation process. The solid content of the collected grape increased to 60% through the mechanical dewatering test. Drying of grape marc up to 20% is required for stable storage and keeps the water activity as low as possible. The higher heating value of grape is 17.50 MJ/kg oven dry, which was determined from ultimate analysis.

Drying followed by mechanical dewatering could be the possible method to reduce the dried grape marc. This method could be the alternate solution to the drying combustion, drying gasification, and drying pyrolysis. Due to the economic consideration, dried grape marc is converted to valuable products such as animal feed or solid fuel, instead of land spreading. Char, tar, and ash could be produced through combustion or gasification. Tar can be returned to agricultural land because it contains polycyclic aromatic hydrocarbons (PAHs). Pyrolysis is operated at lower temperatures, and the generation of PAHs, biochar, and charcoal from the dried grape marc is generally low. Table 4.3 shows the mapping of various process outputs from repurposing options.

If the combustion of the products is incomplete, the concentration of greenhouse gases such as carbon monoxide and methane will not meet emissions standards. It causes carbon footprint issue. Poor composting practices cause CO_2 emissions. A factor related to the moisture content of the grape marc before thermal processing, increasing the water content annuls all benefits of the thermal processes with coal offsetting compared with composting. It highlights the importance of mechanical dewatering. Gasification is not a favourite option due to its high capital cost and low return on investment.

TABLE 4.3

Mapping of Various Process Outputs from the Repurposing Options

Repurpose	Land Spreading	Composting	Drying			
			Combustion to generate electricity	Gasification to generate combined heat and power	Biochar (pyrolysis)	Charcoal
Grape marc	X	X				
Biochar					X	X
Electricity			X	X		
Heat				X	X	
Ash		X	X			
Effluent		X	X	X	x	X
Gasification char				X		
Gasification Tar				X		

Process outputs

The drying to generate dried grape mars has the best economic outlook if it is sold as solid fuel. If such plants were to convert to biomass boilers with grape marc as part of the feedstock mix, then offsetting yields a very favourable carbon footprint. Pyrolysis to produce charcoal or biochar has a significant shortfall between the revenue needed and that able to be earned (Jones et al., 2020).

Example 2: Sago Waste

The environmental issues caused by the large amount of sago waste produced by sago processing industries are seriously discussed. Those dried residues contain an estimated 58% high starch content and are utilised for other applications. The drying of sago waste using a fluidised bed dryer (FBD) was investigated which offers more advantages than other drying methods. The FBD system is operated with a velocity of 1.3 m/s and an operating temperature of 50°C in the optimum condition. The final moisture content is expected at 10% or a moisture ratio of 0.25 in sago waste. It can be utilised as animal feed, as bacterial growth is prevented, and easy to pack. A laboratory investigation concluded that the velocity range between 1.0 and 2.2 m/s is suitable for the fluidisation and drying of sago waste with a particle size of 2,000 μm for a drying duration of 1 hour. The drying rate and fluidisation profiles are examined and resulted from the excellent fluidisation at particle size range between 500 and 2,000 μm at the air velocity of 1.3 m/s (Rosli et al., 2018).

Example 3: General Food Waste

A study was conducted to develop a high-capacity vacuum-drying food waste disposal system. Figure 4.8 shows the schematic diagram of the system. The study evaluates the economic aspects through the usage of steam and the drying characteristics. A drying system with a capacity of 10,000 kg/h was investigated, and the steam generated from a boiler is used as the drying heating source. The drying cost was evaluated through the investigation in which the water and mixtures (sawdust, flour, and water) were used. Results showed that the average drying speeds were 19.65/h for water and 3.89/h for the mixtures by feeding the steam temperature at 1193.2°C. The drying speed increased when the drying progressed, and it reached a moisture content of 20% (w.b.) or less after drying without a drying delay caused by moisture condensation. The ratio of the drying energy consumption to the input energy supply was 88.9% for water and 99.23% for the mixtures. The total drying period was 5.5 hours for water while recorded at 18.5 hours for mixtures, which were significantly shorter compared with that of the hot-air drying method. The drying speed, water content after drying, drying energy rate, and total drying time were significantly improved compared with those of the case of converting steam to hot air when the boiler steam was directly used as a drying heat source for the disposal of large-volume food waste with high water

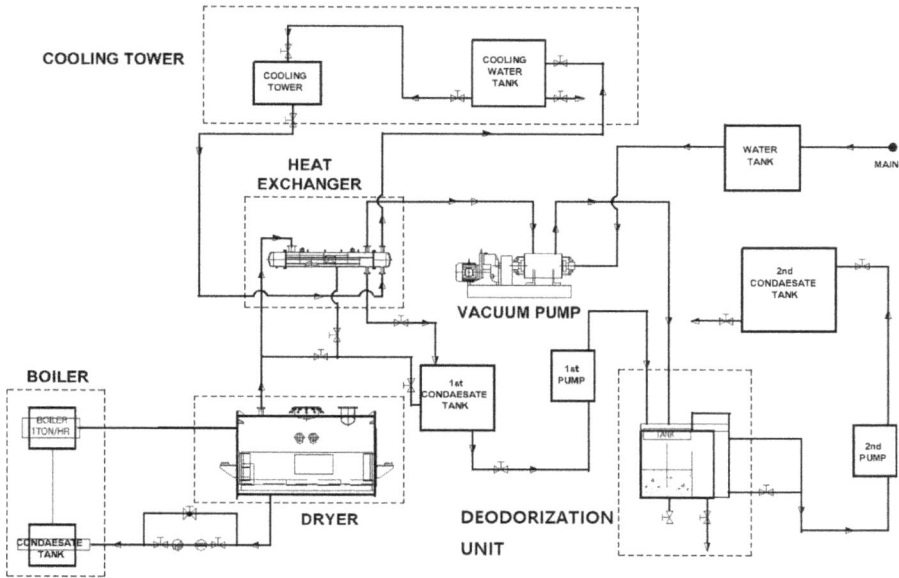

FIGURE 4.8
Process diagram of drying system (NDT Engineering Co. Ltd.) (Song et al., 2020).

content. The drying expenses were lower as compared to those methods that use electricity or gas as a heat source. It indicated that the system can be used for the disposal of large volumes of food waste (Song et al., 2020).

4.3 Organic Crop Residue

4.3.1 Animal Feed from Food Waste

The utilisation of crop residue requires the consideration of precise drying technical aspects for various physicochemical properties, biophysical, and anti-nutritive factors. Despite the widespread utilisation of crop residues in animal feeding, there can be multiple factors that influence the complete conversion of these residues into valuable products suitable for animal feed. For instance, cereal-based crop residues are protein deficient, whereas cell wall as a neutral-detergent fibre can account for as much as 80% of the dry matter, which represent a large source of energy for animals like ruminants. However, the ability of rumen flora-based microorganisms to digest cell wall polysaccharides like hemicelluloses and cellulose is limited by the presence of phenolic and other aromatic compounds. These are generally referred to as lignin. The sorghums are known undergoing for prussic acid poisoning when subjected to high temperatures, which stimulate soybean to produce anti-nutritive factors like inhibitors of trypsin and non-starch polysaccharides. Maize cobs, feeding tubes, and other large particles of food are known to choke due to the blocking of the oesophagus in ruminants, which happens when animals only partly swallow solid particles of food like potato tubers. Considering most of this aforementioned issue, many new techniques have been developed to increase the efficiency of crop residues and by-products utilisation.

Several factors affect the quality of food waste, including:

1. Leaching of nutrients due to storage and processing of food waste.
2. Food waste nutrition or contamination.

Case Study 4.1: Valorising Raw Agriculture Commodities (RAC). Example: Converting Almond or Nuts Edible and Non-Edible Waste into Animal Feed

The raw almond or nut is used to produce roasted and cut almonds or nuts for various confectionery purposes. Nut meats (kernels) in the shell are mechanically shelled. A wet skin waste is generated through the blanching process. The wet skin contains 50% moisture content that is fed to cattle. Dry shell waste with a moisture content of 5–8 wt% is generated and may be used as burning feedstock and fuel or for other commercial non-food uses such as polishing media and biocomposite materials. Figure 4.9 shows the role of drying technologies in the raw agriculture commodities (RAC) process for food wastes like almonds and nuts.

Case Study 4.2: Valorising Raw Agriculture Commodities (RAC). Example: Converting Apple Edible and Non-Edible Waste into Animal Feed

Apple is utilised to produce fruit juices, apple juice concentrate, canned apple slices, and cider vinegar. Expired canned apple slices are possible to process into edible form again after separating the non-edible canned apple slices. The primary food waste of apples is pomace that typically contains 75–85 wt% moisture content, which is mechanically screened and pressed to remove the residual juice. Apple pomace contains a good source of pectin and polyphenols, which are localised in peels. Figure 4.10 shows the main processes such as washing, coring, peeling apples, and coring stems. Peels are removed either mechanically with abrasion or chemically using caustic peeling. Culls and trim material from grading and inspection lines are treated as food waste (Walter et al., 1989).

Case Study 4.3: Valorising Raw Agriculture Commodities (RAC). Example: Converting Berries Edible and Non-Edible Waste into Animal Feed

Berries including strawberries, raspberries, blackberries, boysenberries, gooseberries, blueberries, and cranberries are used to produce canned and

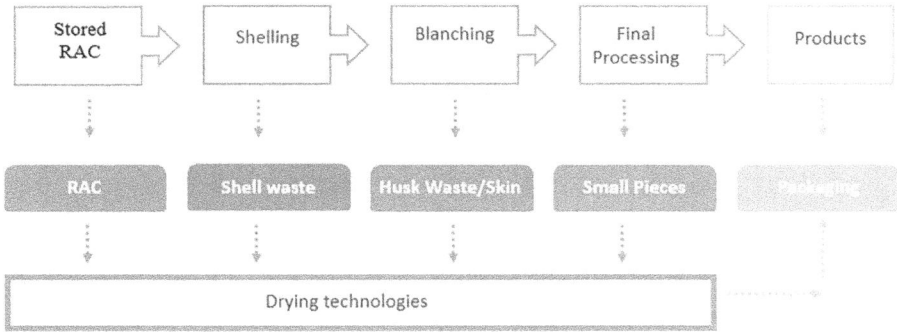

FIGURE 4.9
The role of drying in the process flow diagram of raw agriculture commodities (RAC) processing for food wastes like almond and nuts.

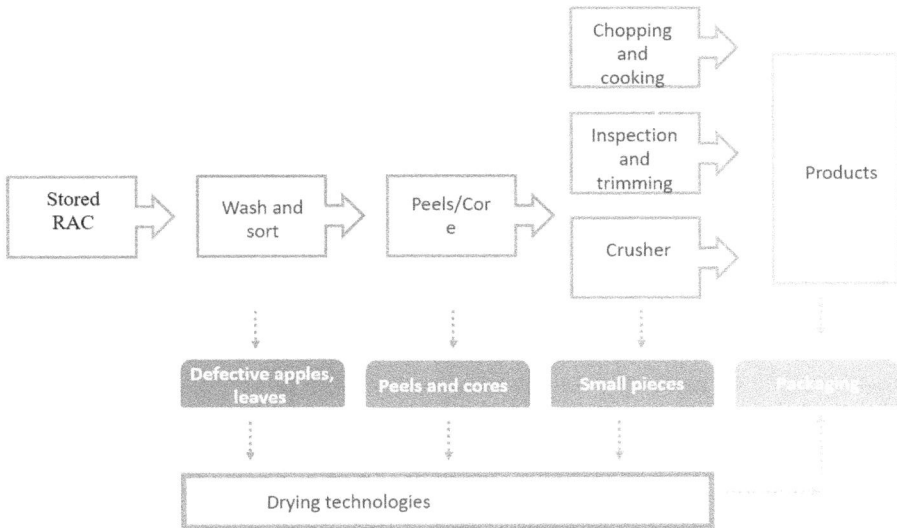

FIGURE 4.10
The role of drying in the process flow diagram of raw agriculture commodities (RAC) processing for food waste like apple.

frozen berries, puree, and wine. The berry wastes are defective berries, stems, and leaves. In berries processing, berries are usually washed using water, dewatered, and frozen in bulk (Figure 4.11). For example, cranberries are then stored frozen and processed year-round. Further to this, it can be processed into a cocktail or berry juice (Walter et al., 1989).

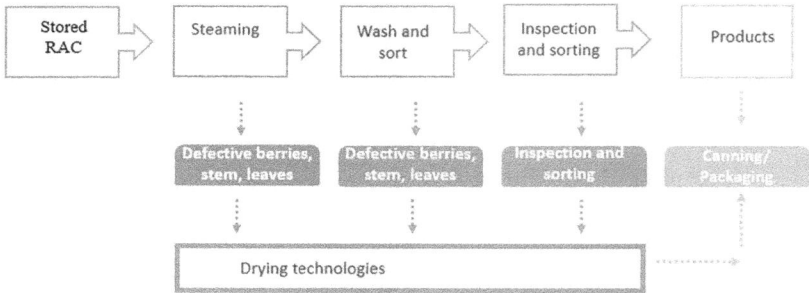

FIGURE 4.11
The role of drying in the process flow diagram of raw agriculture commodities (RAC) processing for food waste like berries.

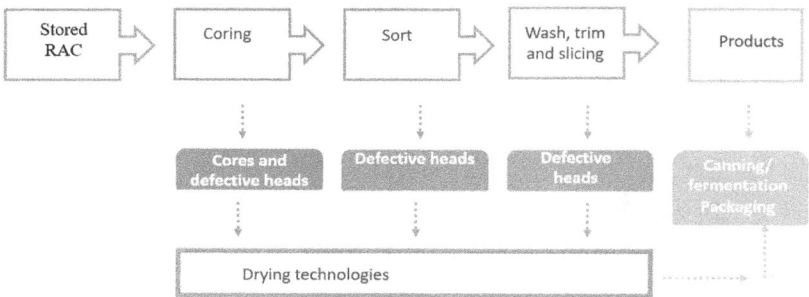

FIGURE 4.12
The role of drying in the process flow diagram of raw agriculture commodities (RAC) processing for food wastes like cabbage or leafy vegetables.

Case Study 4.4: Valorising Raw Agriculture Commodities (RAC). Example: Converting Cabbage or Leafy Vegetables Edible and Non-Edible Waste into Animal Feed

Cabbage or leafy vegetable is canned and sold as uncooked, packaged shredded cabbage or vegetables. Cabbage or leafy vegetable is washed, hand-sorted, and trimmed to remove defective outer leaves (Figure 4.12). The main by-products and wastes are the core, defective head, and leave. Drying technologies can be used to process the disposal of cabbage or leafy vegetables (Walter et al., 1989).

Case Study 4.5: Valorising Raw Agriculture Commodities (RAC). Example: Converting Carrot or Root Vegetables Edible and Non-Edible Waste into Animal Feed and Fuel

Carrot or root vegetable is consumed to produce frozen, canned food, sliced carrots or root vegetable, and mixed vegetable juice (Figure 4.13). The food

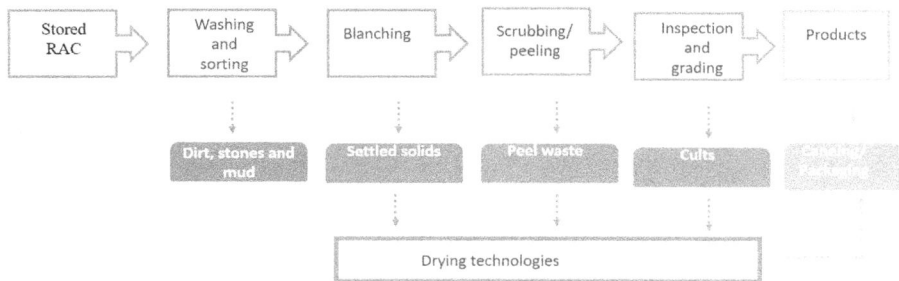

FIGURE 4.13
The role of drying in the process flow diagram of raw agriculture commodities (RAC) processing for food wastes like carrot or root vegetables.

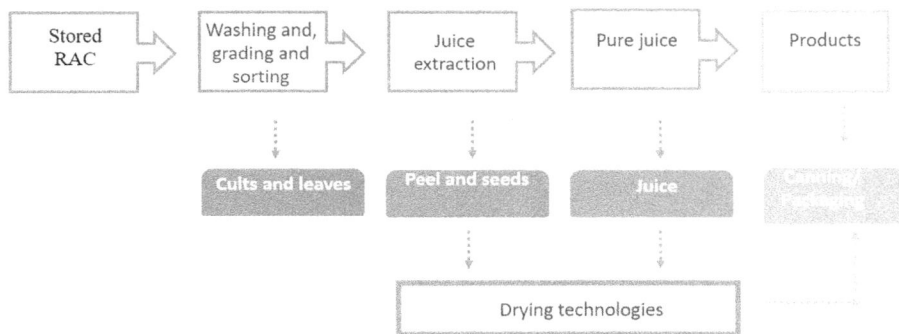

FIGURE 4.14
The role of drying in the process flow diagram of raw agriculture commodities (RAC) processing for food waste like citrus fruit.

wastes are off-grade carrots or root vegetables, crowns, peels, and pulp. Carrot or root vegetable peel/pulp was being burned and used as fuel (Walter et al., 1989).

Case Study 4.6: Valorising Raw Agriculture Commodities (RAC). Example: Converting Citrus-Based Fruit Edible and Non-Edible Waste into Animal Feed

Citrus including oranges, lemons, grapefruit, and limes produce frozen juice concentrate, peel, and essences. Graded whole or halved oranges are individually pressed to remove the juice, which is then further processed before packaging as a frozen concentrate or canned ready-to-drink juice. The by-products and wastes from the pressing device contain the membrane material (pulp), seeds, and residual juice (Figure 4.14). The peel is removed as a separate by-product stream (Chandrasekaran, 2013). The pulp is dried

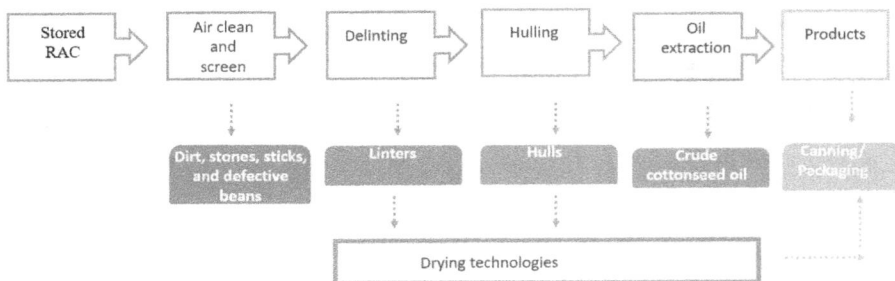

FIGURE 4.15
The role of drying in the process flow diagram of raw agriculture commodities (RAC) processing for food waste like cottonseed.

(11.6 wt% moisture content) before its use as cattle feed. Residual peel waste (53 wt% moisture content) is also fed to cattle (Walter et al., 1989).

Case Study 4.7: Valorising Raw Agriculture Commodities (RAC). Example: Converting Cottonseed Waste into Animal Feed

Cottonseed is utilised to produce crude and refined cottonseed oil. Air scalped is used to remove dirt and other field trash (9 wt% moisture content). The raw seeds are stripped off the remaining fibres. The seeds are then dehulled to remove the tough outer conversion from the seed (Figure 4.15). The dehulled seed is then pressed through a roller mill to produce flakes that can readily be extracted. The extraction is typically by solvent (hexane) although mechanical methods are also used. The food wastes are the hulls and the desolvenised meal, which are generally fed (10 wt% moisture content) (Walter et al., 1989).

Case Study 4.8: Valorising Raw Agriculture Commodities (RAC). Example: Converting Dried Beans, Peas, or Other Beans Edible and Non-Edible Waste into Animal Feed

The dried beans and peas are utilised to produce canned beans, dried beans, and peas. The beans are inspected, cooked, and filled (canned beans) or air-dried (dry packaged beans) (Figure 4.16). By-products and wastes are generated from screening and cleaning operations (Walter et al., 1989).

Case Study 4.9: Valorising Raw Agriculture Commodities (RAC). Example: Converting Dried Grapes Edible and Non-Edible Waste into Animal Feed

Grapes are commonly utilised to produce wine, grape spirits, grape juice, and fruit cocktail. Grape pomace is generated during grape press and comprises peels and seeds. Pomace is generated to produce juice and wine (Figure 4.17).

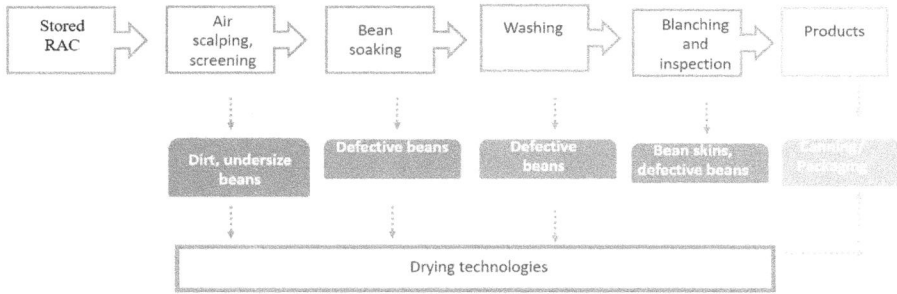

FIGURE 4.16
The role of drying in the process flow diagram of raw agriculture commodities (RAC) processing for food wastes like dried beans, peas, or other beans.

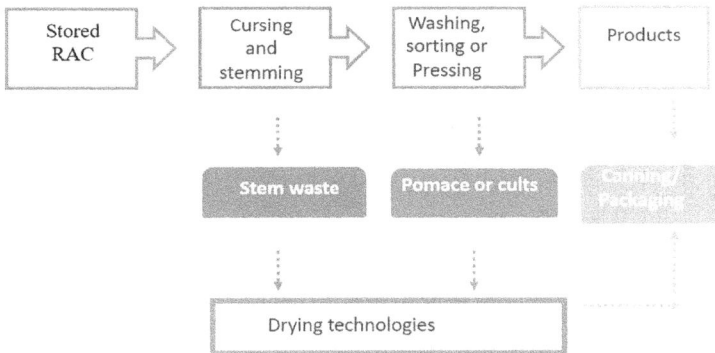

FIGURE 4.17
The role of drying in the process flow diagram of raw agriculture commodities (RAC) processing for food waste like grapes.

It may contain some filter aids such as rice hulls or paper. Semi-solid wastes including tank sludge are produced during grape processing (Walter et al., 1989).

Table 4.4 shows the summary of the existing and potential drying process of various commodities by-products and wastes. It shows that majority of the by-products and wastes are potentially treated through the drying process. By-products and wastes are utilised to produce animal feeds.

4.3.2 Extraction of Valuable Components

RAC bioactive compounds can be extracted using different techniques. The techniques used are supercritical fluid, microwave-assisted, ultrasound-assisted,

TABLE 4.4

Various Drying Techniques Used for the Valorisation of Essential Oil Composition from RAC Food Waste

Drying Technique	Feed Type	Mechanism	Effectiveness and Retained Valuable Compounds	Limitations	Essential Oil (EO) Content	References
Convective drying	Solid fruits, vegetable pomace, vegetables	Moisture exchange between the hot air flowing through the drying chamber and food product.	Longer shelf life, ease of operation, cost-effective, and simple design. Higher yield of EO compounds like Camphene (0.89%), α-Phellandrene (2.50%)), 1,8-Cineole (62.14%), E-β-Ocimene (0.26%), cis-Carveol (0.16%)), Eugenol (0.09%), Sesquiterpene (11.45%).	A high temperature of inlet gas or very dry gas, longer drying time, oxidation; remove off flavours; formation of crust on the product surface because of high temperatures. Lower yields of Sabinene 1%, Linalool 2.77%, α-Humulene 0.11%, α-Terpinyl acetate 0.47%, Eugenol 0.09%, Thymol 1.88%, Monoterpene hydrocarbons 5.91%.	Higher yield of essential oil than during microwave drying of herbs (rosemary and basil).	Ahmed et al. (2011), Barbosa et al. (2015), Raghavi et al. (2018), Saavedra et al. (2017), and D. B. Singh and Kingsly (2008)
Spray drying	Liquid—i.e. puree, solutions, juices, milk, vegetables	Conversion of liquid product into dry powder form.	Lower levels of moisture content and higher quality products, longer shelf life; similar shape and size of dried material, constant operation, cost-effective. Lemon grass EO at high encapsulation efficiency (84.75%) and oil retention (89.31%).	Might lead to loss of bioactive compounds and the high temperature, equipment size, products, stickiness with large fat content that needs to be reduced, higher equipment-installation cost	-	Anandharamakrishnan and Ishwarya (2015), Desobry et al. (1997), Lipan et al. (2021), and Saavedra et al. (2017)

(Continued)

TABLE 4.4 (*Continued*)
Various Drying Techniques Used for the Valorisation of Essential Oil Composition from RAC Food Waste

Drying Technique	Feed Type	Mechanism	Effectiveness and Retained Valuable Compounds	Limitations	Essential Oil (EO) Content	References
Freeze drying	All types of food waste	Dual steps process: (1) freezing the water available from the raw material; (2) heating of the frozen solid to make the moisture undergo sublimation.	Prevents damages due to oxidation Minimises the loss of chemical compounds Minimal shift of soluble solids and shrinkage, retains, volatile compounds, maintains the porous structure. • Higher yields of EO compounds like α-Thujene 0.25%, p-Cymene 5.29%, Cis-α-Bisabolene 5.95%, Caryophyllene oxide 2.54%, β-Myrcene 0.79%	High cost, time-consuming, and expensive process Lower yields of EO compounds like β-Pinene 0.11%, Borneol 0.62%, Terpine-4-ol 0.33%, Thymol 61.86%, Carvacrol 5.47%, β-Caryophyllene 4.73%, α-Humulene 0.19%, Ledene 0.2%, Spathulenol 0.82%	Preservation of most EOs	Lipan et al. (2020) and Rahman (2020)
Osmotic dehydration	Fruits and vegetable waste	Reduces moisture by immersing the raw material in a high osmotic pressure solution → transfer of moisture from the food to the solution driven by the difference in osmotic pressure. Increased yield of rosmarinic acid 9.847 mg/100 g	Maintaining sensory parameters and physicochemical properties Enhancing product quality. Reduced yield of caffeic acid 4.382 mg/100 g and chlorogenic acid 179 mg/100 g	High content of sugar/salt in the product when dehydrated and high final moisture content which require further drying, difficulty in predicting chemical content.	-	Bellary et al. (2011), de Mendonça et al. (2017), Pisalkar et al. (2011), and Szychowski et al. (2018)

(*Continued*)

TABLE 4.4 (Continued)

Various Drying Techniques Used for the Valorisation of Essential Oil Composition from RAC Food Waste

Drying Technique	Feed Type	Mechanism	Effectiveness and Retained Valuable Compounds	Limitations	Essential Oil (EO) Content	References
Intermittent drying	Fruits and vegetables wastes	Intermittent microwave heating assisted applying microwave energy as sequential pulses during the drying process	Preserve bioactive compounds, texture, colour, and reduce the browning effects and extend the shelf life	Higher power ratio can damage certain important volatile compounds such as ascorbic acid	-	Pham et al. (2018), Pisalkar et al. (2011), and Ramya and Jain (2017)
Microwave drying, Vacuum-microwave drying, combined convective drying followed by vacuum-microwave drying		Microwave heating assisted applying microwave energy	Better preservation of colour than convective drying Higher yield of EO compounds like Sabinene 3.12%, β-Copaene 0.24%, α-Terpinyl acetate 0.67%, Geraniol 0.68%, Methyl eugenol 8.61%, Eugenol 4.45%, Monoterpene hydrocarbons 6.46%	High loss of volatiles Lower yield of EO compounds like 1,8-Cineole 56.78%, β-Phellandrene 0.23%, γ-Terpinene 0.22%, E-β-Ocimene 0.10%, Z-Sabinene hydrate 0.60%, Borneol 5.88%, Cinnamyl alcohol 0.10%	Increased EO yield in thyme, oregano, and rosemary	Calín-Sánchez et al. (2020) and Pham et al. (2020)

pressurised liquid, pulsed electric, enzyme-assisted, and high-voltage electrical discharge or electrohydrodynamic (EHD). Detailed challenges, opportunities, and recommended compounds are shown in Table 4.5.

4.3.3 Biofuel Production

Figures 4.9–4.17 show the role of drying technologies in the process flow diagram of raw agriculture commodities (RAC) processing for different food wastes. On top of converting it to animal feeds, the food waste can be processed into biofuel via different routes such as liquefaction, gasification, pyrolysis, transesterification, and fermentation (Table 4.6).

4.4 Catering Waste

Catering wastes are suitable to direct processes into biofuel after drying, land spreading, and composting. There are several drying technologies available for processing catering food waste into fuel. For example, sun drying, solar drying, bin drying, vertical continuous drying, cylindrical drying, funnel cylindrical drying, continuous mobile, and steady tray drying, batch drum drying, and continuous belt conveyor drying of catering waste. These different drying method applications depend on the initial and desired final moisture content of materials dried and the availability and feasibility of the drying technologies. Table 4.7 shows different drying technologies used in the drying of catering waste.

Sun drying requires high amount of sunlight and large drying area. This method is inexpensive, yields uniform drying for food waste, and was proven to give greater yields of β-eudesmol (30.5%), α-ocimene (16.6%), and nerolidol (10.7%) from the dried food waste products. However, a proper measure must be taken to avoid the risk of contamination in the dried products. An alternative to open sun drying, solar drying can be performed during low level of irradiance. Solar drying also significantly reduces the drying time of food waste by 24% compared to open sun drying to achieve 18% moisture content. However, the method requires solar radiation equipment, which increases the cost of drying.

Bin drying is a method whereby the drying of material is conducted in a large cylindrical or rectangular bin with a wire mesh bottom that is filled with hot air. Bin drying is a good option for drying low-moisture content food wastes such as grains, fruits, and vegetables. This method yields good final dried product quality but normally can be only carried out at a low capacity and could cause high damage to the product due to over-drying.

TABLE 4.5

Extraction of Valuable Components from RAC Using Different Extraction Techniques

Extraction Procedure	Effectives and Retained Bioactive Compounds	Limitations	Recommended Compounds	References
Supercritical fluid extraction	• High extraction yield • Faster extraction rate • Absence of toxic residue and solvent and toxic in the final product • Recyclability and reuse of supercritical fluid • Carbon dioxide is the most used supercritical solvent, which is non-corrosive inert, non-flammable, nontoxic and safe for humans, animals and the environment. • Preservation of the bioactive compounds from oxidation and high temperature. • Combined with pulsed electric fields or ultrasound can lower processing cost and increase efficiency	• Carbon dioxide is a nonpolar compound and, therefore not suitable for extraction of polar molecules. • Not suitable for a large amount of pharmaceutical and drug samples • High-cost system and thermodynamically complicated	Flavonoids, polyphenols, carotenoids, phytosterols, fatty acids, sugars, and volatile compounds	da Silva et al. (2016), Jiménez-Moreno et al. (2020), Pereira and Meireles (2010), Rodríguez García and Raghavan (2022), Sagar et al. (2018), and Uwineza and Waśkiewicz (2020)

(Continued)

TABLE 4.5 (*Continued*)

Extraction of Valuable Components from RAC Using Different Extraction Techniques

Extraction Procedure	Effectives and Retained Bioactive Compounds	Limitations	Recommended Compounds	References
Microwave-assisted extraction	• Faster extraction process • Higher extraction yield • Low solvent consumption • Human and eco-friendly technique • Improved quality and high selectivity of desired extracts • Lower cost • Ease of operation • Lower energy requirement	• Expensive equipment at an industrial scale • Not recommended for nonpolar or heat-sensitive compounds • Necessity of optimisation of method and parameter to effectively extract the compounds of interest and careful controlling	For fast extraction of phenolics, carotenoids, pectin, oils, polyphenols, flavonoids and proteins Total carotenoid extraction yield at 360 W increased rapidly from 156 to 236 mg/100g during 4 minutes of extraction time	Azmir et al. (2013), Bagade and Patil (2021), Jiménez-Moreno et al. (2020), Mohan et al. (2020), and Rodríguez García and Raghavan (2022)
Ultrasound-assisted extraction	• Higher extraction yield • Reduced extraction time • Lower consumption of energy • Lower chemical usage • Suitable for heat-sensitive ingredients	• Lack of compound selectivity during the extraction • Production of free radicals at higher sonication powers • Higher ultrasonic power can disintegrate the compounds of interest and lower ultrasonic power may result in a low yield. • Generation of heat when increase in power or longer time. • Complex system	Chlorophyll, phenolics, carotenoids, lipids, and polysaccharides anthocyanin increased with increasing reaction time at a temperature (30°C) to 5.63 mg/g, while the yield decreased at 50°C to 5.23 mg/g	Dassoff et al. (2021), Jiménez-Moreno et al. (2020), Mohan et al. (2020), and Rodríguez García and Raghavan (2022)

(Continued)

TABLE 4.5 (Continued)

Extraction of Valuable Components from RAC Using Different Extraction Techniques

Extraction Procedure	Effectives and Retained Bioactive Compounds	Limitations	Recommended Compounds	References
Pressurised liquid extraction	• Lowered extraction time and reduced solvent volumes requirement compared to conventional techniques. • Better than supercritical fluid extraction for polar compounds • Eco-friendly • Improvement in the mass transfer rates and higher extraction yield	• Higher equipment cost • Limited amount of processing sample • Not suitable for processing thermolabile samples • Higher pressure needed. • Equipment maintenance difficulty	Polyphenols and phytochemicals Higher yields of 1,8-cineole (8.1%), linalool (34.1%), linalyl acetate (30.5%), and camphor (7.3%)	Jiménez-Moreno et al. (2020), Mohan et al. (2020), and Rodríguez García and Raghavan (2022)
Pulsed electric field	• Higher extraction yield • Shorter extraction time • Lower energy consumption • Operations at low temperature and, thus, preserves of heat-sensitive compounds • Suitable for large-scale applications • Eco-friendly	• The system is less stable and safety issues. • Higher maintenance cost • Labour-intensive	Anthocyanins, β-carotene, phytosterols, polyphenols Higher yields of vanillin (+14%), theobromine (+25%), caffeine (+34%), linalool (+114%), and limonene (+33%)	Jiménez-Moreno et al. (2020), Mohan et al. (2020), and Rodríguez García and Raghavan (2022)

(Continued)

TABLE 4.5 (*Continued*)

Extraction of Valuable Components from RAC Using Different Extraction Techniques

Extraction Procedure	Effectives and Retained Bioactive Compounds	Limitations	Recommended Compounds	References
Enzyme-assisted extraction	• Appropriate to extract bound compounds. • Higher extraction rate and yield • Higher reaction specificity • Eco-friendly and lower energy consumption • Improves the quality of the extract	• Higher cost of enzymes required for large sample volumes. • Not suitable at industrial scale due to sensitivity and complex behaviour of enzymes • Necessity for calibration and optimisation of process parameters such as temperature, pH, moisture content, time, and enzyme concentration for each bioactive compound and specific process	Bounded phytochemicals, carotenoids (especially phenolics, lycopene), and oils Eugenol yield extracted per enzyme: Lignocellulase 64.29% Amylase 70.22% Cellulase 62.24% Pectinase 63.33%	Jiménez-Moreno et al. (2020), Mohan et al. (2020), and Rodríguez García and Raghavan (2022)
High-voltage electrical discharge or electrohydrodynamic (EHD)	• Lower energy consumption • Higher extraction efficiency • Lower solvent requirement • Shorter extraction time • Lower extraction temperature and, thus, recommended for heat-sensitive compounds. • Higher quality of extracted ingredients	• Lesser selectivity in comparison with pulsed electric field • System feasibility at/or industrial or pilot levels is not known	Flavonoids, polyphenols, phenolic compounds, bioactive compounds, and β-carotene Yields of compounds extractable in ng/mL: Diosmetin 63.431, Hydroxytyrosol 37.787, Luteolin 306.348, Oleanolic Acid 1,044.092, Quercetin 65.791, Rosmarinic Acid 5,510.994, p-Cymene 0.078, Camphor 2.328. Certain compounds like Thymol and Carvacrol could not be extracted	Dastjerdi et al. (2022), Deng et al. (2018), Maher et al. (2020), Salehi and Taghian Dinani (2020), Shahram and Taghian Dinani (2019), and Yan et al. (2017)

TABLE 4.6

Examples of Additional Valorisation Routes After Drying to Produce Biofuel Production from Food Waste

Routes	Drying/Pre-Treatment Method	Setting	Type of Feedstock	Type of Biofuel	Yield /Energy Recovery/ efficiency	Country	Reference
Liquefaction	Air drying, chopping, sieving, desiccating in vacuum	A temperature between 150°C and 200°C is required, 8 MPa	Dairy waste	Biocrude oil	34.44%	China	Lu et al. (2017)
Gasification	Chopping and sieving	Optimal gasification requires dry fuels of uniform size, with moisture content no higher than 15%–20%	Coffee husk	Fuel gas	7.76 (MJ/N/M3)	Brazil	de Oliveira et al. (2018)
			Rice straw	Fuel gas	33.78%	China	Liu et al. (2018)
			Spent coffee grounds loaded with cobalt	Syngas (H$_2$ and CO)	H$_2$-1.6 mol%; CO-4.7 mol%	Korea	Cho et al. (2018)
Pyrolysis	Grinding and sieving, squeezing, air drying, oven drying	A temperature between 200°C and 520°C is required	Greenhouse	Biochar	40.22 wt%	Turkey	Merdun and Sezgin (2018)
			Vegetable wastes	Bio-oil	30.4 wt%	Turkey	Merdun and Sezgin (2018)
			Sugarcane residues	Bio-oil	From leaves-52.5 wt%	Thailand	Pattiya and Suttibak (2017)

(Continued)

TABLE 4.6 (*Continued*)

Examples of Additional Valorisation Routes After Drying to Produce Biofuel Production from Food Waste

Routes	Drying/Pre-Treatment Method	Setting	Type of Feedstock	Type of Biofuel	Yield /Energy Recovery/ efficiency	Country	Reference
			Sugarcane leaves and tops	Bio-oil	From tops-59.0 wt%	Thailand	Pattiya and Suttibak (2017)
Transesterification	Microwave drying	The complete transesterification reaction was achieved in 4 h with 1:30 M oil/methanol ratio, 5 wt.% catalyst at 65°C	Triacylglycerols	Biodiesel	NR	Luxembourg	Muller et al. (2014)
			Recycled cooking oil	Biodiesel	98.95%	Iran	Tahvildari et al. (2015)
			Fatty acid methyl ester (FAME)	Biodiesel	100%, 98%	Japan, India	Bharat and Bhattacharya (2012) and Farobie and Matsumura (2015)
Fermentation	Air drying		Banana Stem	Bioethanol	25%	Malaysia	Hossain et al. (2019)
			Potato peel waste	Bioethanol	41%	Greece	Hossain et al. (2017)
			Mandarin peel waste (MPW)	Bioethanol	50-60 L/1,000 kg MPW	Spain	Boluda-Aguilar et al. (2010)

Drying and Valorisation of Food Processing Waste

TABLE 4.7

Different Drying Technologies for Catering Waste

Drying Techniques	Advantage	Limitations	Reference
Sun drying	Good final product quality High product uniformity Increase nut splitting Greater yields of β-eudesmol (30.5%), α-ocimene (16.6%), nerolidol (10.7%)	Negative climate condition Risk of contamination	Kashaninejad and Tabil (2009), Midilli (2001), Rostami et al. (2006), and Zanon Costa et al. (2020)
Solar drying	Good final product quality Solar drying can be performed during low level of irradiance. Drying time reduced 24% compared to open sun drying to achieve 18% moisture content Higher yields of cis-β-Farnesene, 19.64%, Bisabolol oxide A 38.36%, En-in-dicycloether 4.73%, Chamazulene 3.36%	Expensive solar radiation equipment Lower yields of 9,12-octadecadienoic acid (Z, Z) 0.19%, trans-caryophyllene 0.18%	
Bin drying	Good final product quality	Low capacity High damage of product	
Vertical continuous drying	Less energy required for operation	Negative effect on nut splitting	
Vertical cylindrical drying		Low product uniformity	
Funnel cylindrical drying	Ability to control movement and holding time	Usage of higher temperature	
Continuous mobile and steady tray dryer	Ability to use different drying temperature High product uniformity	Usage of high amount of fuel	
Batch drum dryer	High product uniformity	Low capacity	
Continuous belt conveyor dryer	Ability to use different drying temperature	Usage of high amount of fuel	

In other drying methods for catering food waste, vertical continuous drying uses relatively less energy for operation compared to the aforementioned methods. However, it could cause a negative effect on nut splitting. Vertical cylindrical drying is known to have low product uniformity, thus using funnel cylindrical drying gives higher quality dried product due to its ability to control movement and holding time of the dried food waste. Continuous mobile and steady tray dryer and batch drum dryer are another drying method which can provide high uniformity in the dried product. In the case that variation of drying temperature is needed, the continuous belt conveyor dryer and continuous mobile and steady tray dryer can be utilised.

Land spreading of food waste refers to its spreading across the land where organic matter and fertilising elements in food waste are returned to the soil. In terms of land spreading, there are two options: (1) mulching – direct spreading of a layer of food waste on top of the soil, and (2) incorporation of food waste into the soil through methods such as ploughing (Turner et al., 2022). Composting, on the other hand, is the process of biologically degrading heterogeneous solid organic materials, for example, food waste, in controlled moisture to produce a stable material that can be used as organic fertiliser. Composting is a self-heated process and can occur aerobically or anaerobically. Before land spreading and before/after composting process, drying technology may be required to remove excess moisture from the food waste.

A conventional fan can be used for the drying of food compost at room temperature. The highest speed setting was applied to maximise the drying process to lower the moisture content. It also reduces the time that takes to dry the food waste since it has high energy in reducing the excess water content. The moisture content was also reduced per unit time. Apart from reducing moisture, it can also provide a solution to the long duration associated with the drying process. The humidity of the surrounding region also affects the content of moisture of the food compost when using a conventional fan (Rostami et al., 2006).

Convective hot-air-drying technique was applied by setting the temperature at 50°C or above. The duration of oven drying takes about a day. The food compost becomes mostly dried as the method of reducing excess water content is efficient. The heat emitted from the oven will be absorbed by the food compost during the process of drying (Abdullah et al., 2018; Inbar et al., 1989; Mohammed et al., 2017). Compared to conventional fan drying, oven drying offers a faster and more efficient drying of food waste for the purpose of composting (Figure 4.18). This is because during the slow conventional fan-drying process, food waste is exposed to flies and dust which in turn will increase the probability of maggots breeding in the food compost.

On a different note, there are no differences observed in the functional groups available in food waste dried using conventional fan drying and oven drying. The FTIR spectra of all samples showed that the detected compounds corresponding to the wavelength are carboxylic acids (3,500–3,000 cm^{-1}), carbon dioxide (3,000–2,000 cm^{-1}), alkyl aryl ether/tertiary alcohols (1,500–1,000 cm^{-1}), halo compounds (1,000–800 cm^{-1}), nitro compounds (2,000–1,500 cm^{-1}), alkenes

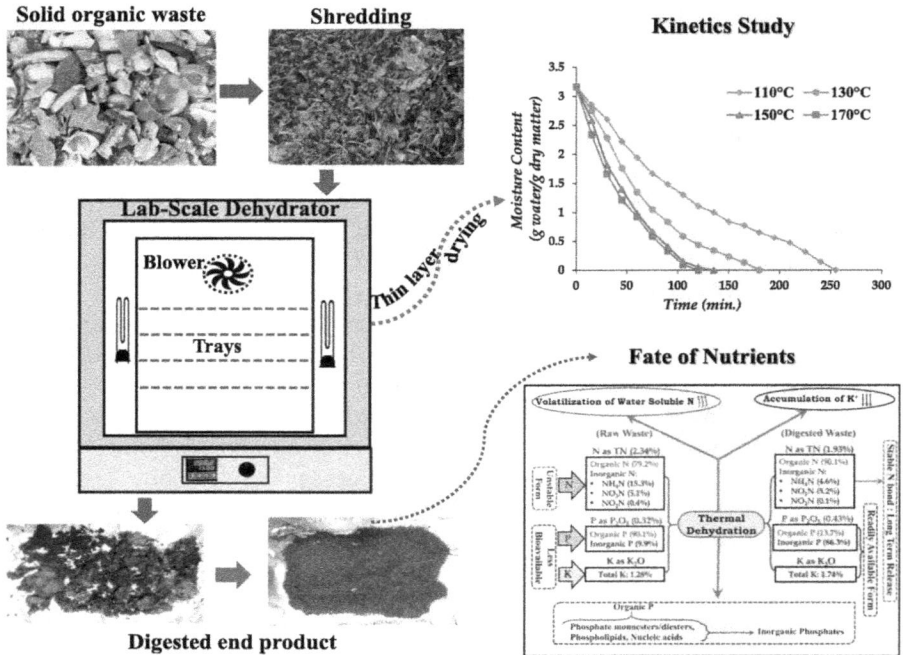

FIGURE 4.18
Summary of dehydration mechanism and fate of nutrients during the thermal digestion of the solid organic waste (Kumar and Gupta, 2022).

(1,000–900 cm^{-1}), and alcohol compounds (3,000–2,500 cm^{-1}) (Abdullah et al., 2018).

4.5 Animal By-Products, Seafood Waste, and Bone Waste

Animal by-products refer to leftover parts from animal origin which is not consumed by human. Rendering process is normally applied to separate reusable parts of animal by-products, and drying is involved as part of the process. For example, high-temperature heating can be used to separate fat from solid animal waste to produce lard and tallow (Hicks & Verbeek, 2016). The objective of the drying process is to separate water during extraction of fat and protein components including blood, bones, and other meat materials from animal by-products. Examples of available drying method include the use of rotary dryer in chicken feather rendering (Woodard & Curran, 2006), and batch cooker and disc dryer/cooker for animal by-product in poultry and red-meat industries (Lim, 2019). Table 4.8 shows several examples of

TABLE 4.8

Examples of Animal By-Product Waste with Their Valuable Active Ingredients and Potential Applications

Types of Animal By-Product	Valuable Active Ingredients/Properties	Function	Application	References
Dairy waste and by-products (sweet whey, acid whey, whey protein concentrate)	Good water stability and excellent oxygen barrier properties	Film-forming capabilities	Raw material for transparent, colourless, and edible bioplastic	Peydayesh et al. (2022)
Fish gelatine-processing waste	Gelatine (e.g.: Gelatine extract of big eye tuna skin had glycine, up to 32% of total amino acids, and hydroxyproline (proline and alanine))	Biomedical properties	Food, pet food, and biomedical industry	Aleksanian et al. (2022)
Fish by-products	Fish protein hydrolysate	Biomedical properties	Food and biomedical industry; fertiliser	Pepi (2022)
Gelatine-digested sludge	–	–	Easy handling of sludge final disposal	Scalcon et al. (2018)
Whey protein concentrate	Essential amino acids, bioactive peptides, and anti-oxidants	Promote overall health	Incorporation in nutritious food products	Patel (2015)
Meat processing waste	Protein, collagen and gelatine	Feedstock for collagen and gelatine production	Food, pet food and biomedical industry	Hicks and Verbeek (2016)
Seafood waste	Proteins, essential fatty acids (long chain n–3 polyunsaturated fatty acids (omega-3 PUFA), mainly eicosa pentaenoic acid (EPA, C20:5; n–3) and docosa hexaenoic acid, (DHA, C22:6; n–3)), vitamins, minerals	Feedstock for collagen and gelatine production	Food and biomedical industry	Djagny et al. (2001)
Salmon bone	Calcium rich (39.15%), lipid content (22.98%), protein (14.82%)	Improve bone and joint health	Calcium-rich food products	Hirunrattana and Limpisophon (2019)

animal by-product waste with their valuable active ingredients and potential applications.

In a study of fish gelatine and its waste drying using spray dryer, it can be concluded that any design of spray dryer should be done based on technological conditions of the drying process including the type of material dried, temperature resistance, viscosity of the liquid medium, type of spray unit, and steam-generation speed (Aleksanian et al., 2022). The leftover parts of fish unused for gelatine production can be further processed as value-added materials in nutritious food, pet food, and biomedical industry.

In the case of gelatine-processing waste, like sludge, drying technology is required not for any food industry applications but to serve other objectives. For example, gelatine sludge contains 98% water, and thus drying is an important process to effectively reduce the sludge volume and handling, transport, and storage costs (Scalcon et al., 2018). The use of hot-air drying (HAD) at 80°C, 110°C, and 140°C, mass flow rate of 0.5 kg/min for drying digested sludge derived from gelatine processing, shows that the sludge apparent density and porosity vary to drying rates and critical moisture content. The shrinkage during drying was evidenced by the decreasing thickness, volume dimensions, and altered surface morphology. However, porosity content of dried sludge is independent of drying temperatures.

Dairy industry such as cheese production involves fats and caseins extraction from milk, whereby 80%–90% of the processed milk volume becomes liquid whey, a dairy by-product. Dairy waste and by-products, namely, the sweet whey, acid whey, whey protein concentrate, have good water stability and excellent oxygen barrier properties which contribute to its high film-forming capabilities. Thus, drying can be employed as a part of the process for preparing this waste as raw material for transparent, colourless, and edible bioplastic (Peydayesh et al., 2022). Freeze drying has also been employed after the lactose and protein separation from whey to preserve the quality of both protein and lactose in the dried product (Das et al., 2016).

Seafood waste is the leftover of seafood processing including shell, head, bones intestine, fin, skin, voluminous amounts of wastewater discharged as effluents, and low-value under-utilised fish (Venugopal, 2021). Studies on more than 40 types of seafood processing by-products showed that these wastes contain 60% proteins, 19% fat, and 22% ash in average (Venugopal, 2021), which indicates their potential for utilisation of downstream products. Seafood waste is also rich in collagen, a valuable material for the food and biomedical industry (Djagny et al., 2001).

Several drying technologies have been employed on seafood waste drying for utilisation. Drying of salmon bones as material for calcium-rich snack preparation, for example, can be realised through boiling by gas retort followed by hot-air drying (Hirunrattana & Limpisophon, 2019). Drying at 180°C for 15 minutes resulted in salmon bone snack with crispness (84 count peaks), hardness (13.16 N), moisture content (2.96%), and calcium (19.23 g calcium per

100 g) (Hirunrattana & Limpisophon, 2019). In fish by-products (fish skins, heads, muscle, viscera, bone, frames, and roe), processing for obtaining fish protein hydrolysate, spray dryers, vacuum freeze dryers, and roller drum dryers can be applied to remove moisture from the raw materials and the product (Pepi, 2022).

4.6 Fruits and Vegetables Waste

Table 4.9 shows examples of fruit and vegetable wastes with their valuable active ingredients and potential applications. In general, peels, seeds, pomace, leaf, rind, and waste grains contain a tremendous amount of active ingredients. Thus, they give nutritional values as functional food products (Ciccoritti et al., 2021; Harada-Padermo et al., 2020), supplements, cosmetics (Pasten et al., 2019; Uribe et al., 2013), and even as medicine (Serrano-Díaz et al., 2013).

Food wastes are generated in vast amounts annually and thus can be considered valuable resources for the recovery of valuable components for countless applications. The availability of physical, chemical, and biological recovery technologies enables target compound recovery and recycling for edible and non-edible purposes. Detailed info related to bioactive compounds can be referred to in Chapter 2. For example, recovery of anti-oxidants from food waste is beneficial for pharmaceutical applications, rich fibrous content in food waste can be utilised in functional food products, and sludge from plant-based oil processing can be used as fuel. In this section, different drying methods will be discussed in the context of the recovery of valuable components in food waste.

Hot-air drying (HAD) is one of the technologies available for recovery of anti-oxidants in food waste. It is interesting to note the differences between anti-oxidant activity of HAD products in comparison with other drying methods. For instance, HAD results in product with higher anti-oxidant activity compared to microwave-dried product from papaya tissue (Nieto-Calvache et al., 2019). On the other hand, the anti-oxidant properties are lower when compared to freeze-dried products from mango peels (Sogi et al., 2013).

Freeze drying combines low-temperature drying with low pressure to remove moisture from food samples via sublimation. Freeze drying can be used to preserve important nutritional qualities in food waste, yielding a dried product of high quality (Valadez-Carmona et al., 2017). For example, low-temperature drying via freeze drying can be used for the recovery of anthocyanins and flavanols from saffron residues because they degrade at 110°C and 125°C (Serrano-Díaz et al., 2013). However, freeze drying requires a long drying period and is sometimes not feasible due to the high cost of freeze dryer.

TABLE 4.9

Examples of Fruits and Vegetable Waste with Their Valuable Active Ingredients and Potential Applications

Types of Waste	Valuable Active Ingredients	Function	Application	References
Iceberg salad fresh-cut processing waste	Fibre and anti-oxidant compounds. Air drying produced flour rich in fibre (>260 g/kg) and polyphenols (3.05 mg GAE/gdw) with anti-oxidant activity (6.04 OD–3/min/gdw)	Digestive health, anti-oxidants	Air-dried product: as functional food ingredients Supercritical-dried products: as bulking agents or oil absorbers (to absorb oil spills, edible oils)	Plazzotta et al. (2018)
Pitted Olive Pomace	High fibre, anti-oxidant, and phenolic contents (16% vanillic acid in air-dried pomace)	Biomedical properties	Incorporated into food products to increase their nutritional quality	Sinrod et al. (2019)
Carrot pomace, tomato peels	Carotenoids - Dried carrot pomace has b-carotene (9.87–11.57 mg) and ascorbic acid (13.53–22.95 mg/100 g)	Natural colourants, biomedical properties	Natural colourants and β-carotene supplements (vitamin A)	Tiwari et al. (2022)
Broccoli stalk slices	Fibre, vitamins, iron, and potassium	Promote overall health	As material for food processing	Salina Md Salim et al. (2016)
Bilberry press cake extrudates	Phenolic compounds (58.4–216.1 mg/g), dietary fibre, protein and lipids	Promote overall health	As ingredients in the food industry	Höglund et al. (2018)
Kyoho grape seeds (*Vitis labruscana*)	Vitamins, dietary fibre	Promote overall health	Value-added food product	Sridhar and Charles (2020)
Nopal pad /Cactus pear peel (*O. ficus-indica*)	Protein (5.71%–8.35%), fat (3.3%–3.68%), vitamins, minerals, and fibres	Promote overall health	Food products	Namir et al. (2017) and Rodriguez et al. (2019)
Vegetable wastes (carrot, leek, celery, and cabbage)	Phenols/flavonoid content, anti-oxidants	Biomedical properties	Functional powders to be used as functional food ingredients, colouring, flavouring ingredients, and as preservatives	Bas-Bellver et al. (2020)

(Continued)

TABLE 4.9 (*Continued*)

Examples of Fruits and Vegetable Waste with Their Valuable Active Ingredients and Potential Applications

Types of Waste	Valuable Active Ingredients	Function	Application	References
Watermelon (*Citrullus lanatus*) rind	Vitamin C, dietary fibre, citrulline, potassium vitamin B-6, bioactive compounds (cucurbitacin, triterpenes, sterols, and alkaloid)	Improve overall health, anti-oxidant	Increase nutritional quality in food products	Mohan et al. (2016)
Yellow passion fruit (*Passiflora edulis var. Flavicarpa*) Peel	Bioactive compounds	Prevention of heart disease, cancer	Increase nutritional quality in food products	Duarte et al. (2017)
Shiitake (*Lentinula edodes*) stipes	Umami compounds	Improve taste in food	Production of a powder ingredient	Harada-Padermo et al. (2020)
Papaya tissue (pulp or peel)	Lignin, cellulose, uronic acids, and proteins	Prevent structure alterations or crystallisation in products like baked product/ice-cream	Addition to baked products or ice creams	Nieto-Calvache et al. (2019)
			In oil/water food emulsions, they can provide anti-oxidant activity diminishing rancidity occurrence	
Cacao pod husks (*Theobroma cacao L.*)	Phenolic compounds, natural anti-oxidants	Regulates cellular activities, anti-oxidants	As ingredients in functional foods	Valadez-Carmona et al. (2017)
Mango peels and kernels	Dietary fibres, pectins, and anti-oxidants, phenolics, carotenoids compounds	Improve overall health, anti-oxidant	As ingredients in functional foods	Nagel et al. (2014) and Sogi et al. (2013)
Orange peel	Aromatic flavour and odour components	Improve taste in food	Herbal teas and food products	Bozkir et al. (2021)
Apple peels	Aromatic flavour and odour components	Improve taste in food	Herbal teas and food products	Moussaoui et al. (2021)

In general, microwave drying provides many advantages including a time- and energy-saving process, and the ability to preserve valuable components in the dried products. However, in some situations, combined drying is a better solution for vegetable and fruit waste drying. Combined drying refers to the application of a combination of two or more drying systems (Lamidi et al., 2019) to achieve drying objectives. In some situations, combined drying is necessary to compensate for the limitations of other drying methods.

Referring to Chapter 3, Table 3.2 shows several examples of combined drying technologies used for drying fruits and vegetable waste. For example, using the solar drying method alone throughout the year may not be suitable in inconsistent weather conditions and changes in seasons, which may cause over- or inadequate drying and contaminations of dried materials (Yi et al., 2020). Another example when combined drying is needed is when dealing with materials that are easily deteriorating in quality in a short time like saffron residues, in which freeze drying may be required to preserve nutritional properties (Serrano-Díaz et al., 2013).

Based on the combined drying approaches discussed, it can be concluded that the application of a combined drying system generally results in faster drying, energy saving, and results in high quality of dried products, compared to using only single drying technique.

4.6.1 Land Spreading and Composting

Fruits and vegetables waste composting is an attractive means for turning this waste into biofertilisers via composting. However, one of the challenges in utilising fruit and vegetable waste is their high water content. Thus, drying of fruit and vegetable wastes prior to composting process is needed in some cases whereby the feedstock is highly wet, due to two factors: (1) fruits and vegetables in nature, contain high moisture, and (2) some fruits and vegetable waste come from food preparation and serving, which unavoidably with the mixture of water in the composition. Thus, reducing the mass and volume of the fruit and vegetable waste via drying means reducing the food waste transportation and composting cost. Volume reduction of food waste via drying can be expressed in percentage volume reduction, V_R (%), with the equation:

$$V_R = \left(\left(V_i - V_f\right) / V_f\right) \times 100\%$$

where V_i and V_f refer to the initial and final volumes of food waste, respectively.

An example of application of drying technology prior to composting is as studied by Nenciu et al. (2022). In this case study, conductive drying using ceramic heaters was used to heat water in fruit and vegetable waste prior to composting until it reaches the vapour stage. There are two concurrent processes involved whereby the drying mechanisms involved water removal via (1) transfer of heat for the evaporation of water to the food and (2) transport of the water vapours formed away from the food (Nenciu et al., 2022).

Drying fruit and vegetable waste at a suitable temperature and duration will influence the compost quality. Drying of fruit and vegetable waste reduced the total N content from 2.2% to 1.81% which is attributed to the evaporation of ammoniacal nitrogen in waste. In addition, the availability of P and K increased from 0.38% to 0.43% and 1.47% to 1.75%, respectively, when the temperature was increased from 110°C to 170°C (Kumar & Gupta, 2022).

Kobayashi et al. (2019) and Ma et al. (2018) studied the use of thermally assisted bio-drying for composting food waste. Conventionally, bio-drying depends solely on high microbial activity to generate heat and subsequently reduce the water content in materials through evaporation. However, in thermally assisted drying, the drying process was expedited. The introduction of char as additional material during bio-drying helped to increase the drying rate of the material and increased the decomposition rate (Kobayashi et al., 2019). Experimental results show that a significant increase in decomposition rate occurs (from 0.06 to 0.36 g-CO_2/h) in Run 6, whereby 50 wt% char was added, and from 0.06 to 0.48 g-CO_2/h observed for Run 8 (20 wt% char added) (Kobayashi et al., 2019). A higher decomposition rate means that the food waste is converted to organic material faster compared to drying with a lower decomposition rate. In addition, staged heating acclimation during thermally assisted bio-drying results in excellent thermophilic inoculum with high metabolic activity and microbial consortia.

4.7 Other Food Wastes

In this section, discussion revolves around other food wastes not mentioned in any of the categories discussed in previous sections of this chapter. In previous sections, we have discussed about organic crop residue, catering/ kitchen waste, animal by-products, fruits, and vegetable wastes. Other types of food wastes include the highly fibrous food waste which is not directly generated as crop residue but rather as food-processing by-products, food waste in the form of mixture of organic and fibrous food waste, and excess seaweeds *Sargassum* found along beaches which can be considered as waste.

Table 4.10 provides some examples of other food wastes with their valuable active ingredients and potential applications including recovery of valuable compounds, animal feed, and biofuel production. Certain food waste may not be accepted as food for human consumption but can be processed as animal feed, due to their carbohydrate, protein, and high dietary fibre content (Ghasemi et al., 2018; Singh et al., 2020). The conversion of plant- and animal-based food waste into animal feed should consider the pre-treatment of food waste, moisture content reduction, microbial load reduction, and the inactivation of anti-nutritional factors (Georganas et al., 2020). The application of food waste as biofuel, on the other hand, is due to the high lignocellulosic

TABLE 4.10

Examples of Other Food Wastes with Their Valuable Active Ingredients and Potential Applications

Types of Food Waste	Valuable Active Ingredients	Function	Application	References
Olive waste cake	Dietary fibre, minerals, fatty acids, tocopherols, phytochemicals (total phenolics, flavonoids and flavanols)	Regulates cellular activities, anti-oxidants	Material for many processing industries (e.g., food and cosmetic)	Pasten et al. (2019), Uribe et al. (2013)
Brewery spent grains (BSG)	Protein and dietary fibre	Building blocks, digestive health	Food supplements/ products, animal feed	Mutlu et al., (2021), Shih et al. (2020), A. P. Singh et al. (2020), and Wallin et al. (2020)
Saffron floral bio-residues	Phenolic/flavanoids compounds Total Polyphenol Index (TPI): 60.82–67.87 Total anthocyanin content (mg D3,5-diG/g dry weight: 39.58–48.44)	Biomedical properties, such as cytotoxic effect against tumour cell lines, antifungal and anti-oxidant	Medicine and ingredient for food product	Ghasemi et al. (2018)
Palletised mixed wastes (including straw and broken kernels of wheat, bread, date palm, grape, pomegranate, and potato and soybean meal)	Fibre	Digestive health	Animal feed	Ghasemi et al. (2018)
Wheat kernel	Bioactive compounds	Prevention of heart disease, cancer	Food supplements and/or in the nutraceutical sector, animal feed	Ciccoritti et al. (2021)

(Continued)

TABLE 4.10 (*Continued*)

Examples of Other Food Wastes with Their Valuable Active Ingredients and Potential Applications

Types of Food Waste	Valuable Active Ingredients	Function	Application	References
Cacao pod husks (*Theobroma cacao L.*)	Phenolic compounds, natural anti-oxidants	Regulates cellular activities, anti-oxidants	As ingredients in functional foods	Valadez-Carmona et al. (2017)
Annatto grains and annatto waste grains flour	Carbohydrates, protein and dietary fibre	Source of energy and building, fibre for digestive health	Food supplements and/or in the nutraceutical sector	Santos et al. (2014)
Sargassum waste	Biologically active compounds: sulphated polysaccharide, phenolics, plastoquinone, phlorotannins, fucoxanthin, fucoidan, sargaquinoic acid, sargachromenol, steroids, terpenoids, and flavonoid	Therapeutic potential	Food, fuel, pharmaceutical products, fertiliser and animal feed	Milledge and Harvey (2016)
Sago hampas (sago flour processing residue)	High lignocellulosic materials (66% starch, 14% fibre, 25% lignin)	-	Feedstock for biofuel production	Abdul Aziz et al. (2013)
Palm oil fatty acid distillate	Fatty acids	-	Feedstock for biofuel production	Ngaini et al. (2022)

materials and fatty acids (Abdul Aziz et al., 2013; Awg-Adeni et al., 2010; Ngaini et al., 2022).

The olive waste cake is a by-product of olive oil processing. It is rich in oleic acid as the predominant fatty acid (63.9% of the total fatty acids). About 84%–87.8% of vitamin E in the fresh and dried olive waste cake consists of α-tocopherol. About 20.7% of the olive waste cake contains phytochemical content, namely, the total phenolics, flavonoids, and flavanols. Studies also showed that drying olive waste cake at 90°C yields 3-hydroxytyrosol as the most abundant phenolic compound (Pasten et al., 2019; Uribe et al., 2013).

Some other food-processing by-products, for example, the brewery spent grain (BSG), saffron floral residues, mixed food waste, kernel, and grains are rich in protein and dietary fibre. Impingement-dried BSG contained the highest protein (18.03 g/100 g dry matter), total phenolic content (2.21 mg GAE/g), radical scavenging activity (1.58 mg AAE/g), total flavonoid content (0.68 mg QE/g), retained lighter colour (L^*, 54.68), and higher total dietary fibre (TDF, 42.40 g/100 g DM), compared to the hot-air-dried BSG (Shih et al., 2020).

4.8 Conclusion

In conclusion, the selection of drying technologies is highly influenced by the food waste characteristics and the final objective of food waste valorisation. Factors such as drying efficiency, nutritional quality preservation, and feasibility should be considered. Regarding the utilisation of food waste as processed food, studies concluded that combined drying is necessary to compensate for the limitations of drying methods in the recovery of valuable components. Important criteria in selecting a pilot- or commercial-scale suitable dryer to valorise food waste need to consider size, density, intended final moisture content, and reaction to heat on the food waste. The other subcriteria to be considered are stated in Chapter 2.

References

Abdul Aziz, S. M., Wahi, R., Ngaini, Z., & Hamdan, S. (2013). Bio-oils from microwave pyrolysis of agricultural wastes. *Fuel Processing Technology, 106*, 744–750. https://doi.org/10.1016/j.fuproc.2012.10.011

Abdullah, M., Rosmadi, H. A., Azman, N. Q. M. K., Sebera, Q. U., Puteh, M. H., Muhamad, A., & Zaiton, S. N. A. (2018). Effective drying method in the utilization of food waste into compost materials using effective microbe (EM). *AIP Conference Proceedings, 2030*, 020120. https://doi.org/10.1063/1.5066761

Ahmed, M., Akter, M. S., & Eun, J. B. (2011). Optimisation of drying conditions for the extraction of β-carotene, phenolic and ascorbic acid content from yellow-fleshed sweet potato using response surface methodology. *International Journal of Food Science & Technology, 46*, 1356–1362.

Aleksanian, Y., Maksimenko, Y. A., Iakubova, O. S., & Bekesheva, A. A. (2022). Study of drying processes of gelatin from fish raw material Study of drying processes of gelatin from fish raw material. *IOP Conference Series: Earth and Environmental Science, 1052*, 012081. https://doi.org/10.1088/1755-1315/1052/1/012081

Anandharamakrishnan, C., & Ishwarya, S. P. (2015). *Spray Drying Techniques for Food Ingredient Encapsulation*. John Wiley & Sons, Ltd. https://doi.org/10.1002/9781118863985

Awg-Adeni, D. S., Abd-Aziz, S., Bujang, K., & Hassan, M. A. (2010). Bioconversion of sago residue into value added products. *African Journal of Biotechnology, 9*(14), 2016–2021.

Azmir, J., Zaidul, I. S. M., Rahman, M. M., Sharif, K. M., Mohamed, A., Sahena, F., Jahurul, M. H. A., Ghafoor, K., Norulaini, N. A. N., & Omar, A. K. M. (2013). Techniques for extraction of bioactive compounds from plant materials: A review. *Journal of Food Engineering, 117*(4). https://doi.org/10.1016/j.jfoodeng.2013.01.014

Bagade, S. B., & Patil, M. (2021). Recent advances in microwave assisted extraction of bioactive compounds from complex herbal samples: A review. *Critical Reviews in Analytical Chemistry, 51*(2). https://doi.org/10.1080/10408347.2019.1686966

Barbosa, J., Borges, S., Amorim, M., Pereira, M. J. V., Oliveira, A., Pintado, M., & Teixeira, P. (2015). Comparison of spray drying, freeze drying and convective hot air drying for the production of a probiotic orange powder. *Journal of Functional Foods, 17*, 340–351.

Barnana. (2023). *Our Story*. https://barnana.com/pages/our-story

Bas-Bellver, C., Barrera, C., Betoret, N., & Seguí, L. (2020). Turning agri-food cooperative vegetable residues into functional powdered ingredients for the food industry. *Sustainability (Switzerland), 12*(4). https://doi.org/10.3390/su12041284

Bellary, A. N., Sowbhagya, H. B., & Rastogi, N. K. (2011). Osmotic dehydration assisted impregnation of curcuminoids in coconut slices. *Journal of Food Engineering, 105*(3). https://doi.org/10.1016/j.jfoodeng.2011.03.002

Bharat, K. T., & Bhattacharya, A. (2012). The production and analysis of biodiesel from waste chicken skin and pork skin fat and a comparison of fuel properties to petroleum derived diesel fuel. *International Journal of Engineering Research and Development, 2*(3). www.ijerd.com

Boluda-Aguilar, M., García-Vidal, L., del P. González-Castañeda, F., & López-Gómez, A. (2010). Mandarin peel wastes pretreatment with steam explosion for bio-ethanol production. *Bioresource Technology, 101*(10). https://doi.org/10.1016/j.biortech.2009.12.063

Bozkir, H., Tekgül, Y., & Erten, E. S. (2021). Effects of tray drying, vacuum infrared drying, and vacuum microwave drying techniques on quality characteristics and aroma profile of orange peels. *Journal of Food Process Engineering, 44*(1). https://doi.org/10.1111/jfpe.13611

Burge, E. (2015). *Nestlé USA Announces That All 23 Factories Achieve Zero Waste to Landfill*. https://www.nestleusa.com/media/pressreleases/nestlé-usa-announces-that-all-23-factories-achieve-zero-waste-to-landfill

Calín-Sánchez, Á., Lipan, L., Cano-Lamadrid, M., Kharaghani, A., Masztalerz, K., Carbonell-Barrachina, Á. A., & Figiel, A. (2020). Comparison of traditional and novel drying techniques and its effect on quality of fruits, vegetables and aromatic herbs. *Foods, 9*(9). https://doi.org/10.3390/foods9091261

Chandrasekaran, M. (2013). *By-Products, Valorization of Food Processing*. CRC Press.

Cho, D.-W., Tsang, D. C. W., Kim, S., Kwon, E. E., Kwon, G., & Song, H. (2018). Thermochemical conversion of cobalt-loaded spent coffee grounds for production of energy resource and environmental catalyst. *Bioresource Technology, 270*. https://doi.org/10.1016/j.biortech.2018.09.046

Ciccoritti, R., Taddei, F., Gazza, L., & Nocente, F. (2021). Influence of kernel thermal pre-treatments on 5-n-alkylresorcinols, polyphenols and antioxidant activity of durum and einkorn wheat. *European Food Research and Technology, 247*(2), 353–362. https://doi.org/10.1007/s00217-020-03627-4

Clancy, H. (2013). *Pepsico recycles snack food waste into energy and fertilizer*. https://www.greenbiz.com/article/pepsico-recycles-snack-food-waste-energy-and-fertilizer

da Silva, R. P. F. F., Rocha-Santos, T. A. P., & Duarte, A. C. (2016). Supercritical fluid extraction of bioactive compounds. *TrAC Trends in Analytical Chemistry, 76*. https://doi.org/10.1016/j.trac.2015.11.013

Das, B., Sarkar, S., Sarkar, A., Bhattacharjee, S., & Bhattacharjee, C. (2016). Recovery of whey proteins and lactose from dairy waste: A step towards green waste management. *Process Safety and Environmental Protection, 101*, 27–33. https://doi.org/10.1016/jpsep.2015.05.006

Dassoff, E. S., Guo, J. X., Liu, Y., Wang, S. C., & Li, Y. O. (2021). Potential development of non-synthetic food additives from orange processing by-products—A review. *Food Quality and Safety, 5*. https://doi.org/10.1093/fqsafe/fyaa035

Dastjerdi, Z. H., Nourani, M., & Dinani, S. T. (2022). Effect of electrohydrodynamic and ultrasonic pretreatments on the extraction of bioactive compounds from Melissa officinalis. *Journal of Food Measurement and Characterization, 16*(1). https://doi.org/10.1007/s11694-021-01183-3

de Mendonça, K. S., Corrêa, J. L. G., de J. Junqueira, J. R., Cirillo, M. A., Figueira, F. V., & Carvalho, E. E. N. (2017). Influences of convective and vacuum drying on the quality attributes of osmo-dried pequi (Caryocar brasiliense Camb.) slices. *Food Chemistry, 224*. https://doi.org/10.1016/j.foodchem.2016.12.051

de Oliveira, J. L., da Silva, J. N., Martins, M. A., Pereira, E. G., & da Conceição Trindade Bezerra e Oliveira, M. (2018). Gasification of waste from coffee and eucalyptus production as an alternative source of bioenergy in Brazil. *Sustainable Energy Technologies and Assessments, 27*. https://doi.org/10.1016/j.seta.2018.04.005

Deng, Y., Ju, T., & Xi, J. (2018). Circulating polyphenols extraction system with high-voltage electrical discharge: Design and performance evaluation. *ACS Sustainable Chemistry & Engineering, 6*(11). https://doi.org/10.1021/acssuschemeng.8b03827

Desobry, S. A., Netto, F. M., & Labuza, T. P. (1997). Comparison of spray-drying, drum-drying and freeze-drying for β-carotene encapsulation and preservation. *Journal of Food Science, 62*(6). https://doi.org/10.1111/j.1365-2621.1997.tb12235.x

Djagny, K. B., Wang, Z., & Xu, S. (2001). Gelatin: A valuable protein for food and pharmaceutical industries: Review. *Critical Reviews in Food Science and Nutrition, 41*, 481–492.

Duarte, Y., Chaux, A., Lopez, N., Largo, E., Ramírez, C., Nuñez, H., Simpson, R., & Vega, O. (2017). Effects of blanching and hot air drying conditions on the physicochemical and technological properties of yellow passion fruit (Passiflora edulis Var. Flavicarpa) by-products. *Journal of Food Process Engineering, 40*(3). https://doi.org/10.1111/jfpe.12425

Farobie, O., & Matsumura, Y. (2015). Biodiesel production in supercritical methanol using a novel spiral reactor. *Procedia Environmental Sciences, 28.* https://doi.org/10.1016/j.proenv.2015.07.027

Georganas, A., Giamouri, E., Pappas, A. C., Papadomichelakis, G., Galliou, F., Manios, T., Tsiplakou, E., Fegeros, K., & Zervas, G. (2020). Bioactive compounds in food waste: A review on the transformation of food waste to animal feed. *Foods, 9*(3), 1–18. https://doi.org/10.3390/foods9030291

Ghasemi, A., Chayjan, R. A., & Najafabadi, H. J. (2018). Optimization of granular waste production based on mechanical properties. *Waste Management, 75,* 82–93. https://doi.org/10.1016/j.wasman.2018.02.019

Greenfield, D. (2009, November 7). *Frito-Lay bakes environmental awareness into production processes.* Control Engineering. https://www.controleng.com/articles/-frito-lay-bakes-environmental-awareness-into-production-processes/

Harada-Padermo, S. dos S., Dias-Faceto, L. S., Selani, M. M., Alvim, I. D., Floh, E. I. S., Macedo, A. F., Bogusz, S., Dias, C. T. dos S., Conti-Silva, A. C., & de S. Vieira, T. M. F. (2020). Umami ingredient: Flavor enhancer from shiitake (Lentinula edodes) byproducts. *Food Research International, 137.* https://doi.org/10.1016/j.foodres.2020.109540

Hicks, T. M., & Verbeek, C. J. R. (2016). Meat Industry Protein By-Products: Sources and Characteristics. In Dhillon, G. S. (Ed.), *Protein Byproducts: Transformation from Environmental Burden into Value-Added Products* (pp. 37–61). Elsevier Science.

Hirunrattana, P., & Limpisophon, K. (2019). Production of calcium-rich snack from salmon bone. *Italian Journal of Food Science, 31*(5), 192–197.

Höglund, E., Eliasson, L., Oliveira, G., Almli, V. L., Sozer, N., & Alminger, M. (2018). Effect of drying and extrusion processing on physical and nutritional characteristics of bilberry press cake extrudates. *LWT - Food Science and Technology, 92,* 422–428. https://doi.org/10.1016/j.lwt.2018.02.042

Hossain, N., Haji Zaini, J., & Mahlia, T. M. I. (2017). A review of bioethanol production from plant-based waste biomass by yeast fermentation. *International Journal of Technology, 8*(1). https://doi.org/10.14716/ijtech.v8i1.3948

Hossain, N., Razali, A. N., Mahlia, T. M. I., Chowdhury, T., Chowdhury, H., Ong, H. C., Shamsuddin, A. H., & Silitonga, A. S. (2019). Experimental investigation, techno-economic analysis and environmental impact of bioethanol production from banana stem. *Energies, 12*(20). https://doi.org/10.3390/en12203947

Inbar, Y., Chen, Y., & Hadar, Y. (1989). Solid-state Carbon-13 nuclear magnetic resonance and infrared spectroscopy of composted organic matter. *Soil Science Society of America Journal, 53*(6). https://doi.org/10.2136/sssaj1989.03615995005300060014x

Jiménez-Moreno, N., Esparza, I., Bimbela, F., Gandía, L. M., & Ancín-Azpilicueta, C. (2020). Valorization of selected fruit and vegetable wastes as bioactive compounds: Opportunities and challenges. *Critical Reviews in Environmental Science and Technology, 50*(20). https://doi.org/10.1080/10643389.2019.1694819

Jones, J., McLaren, S., & Chen, Q. (2020). *Repurposing Grape Marc in Marlborough: The Way Forward from Assessment of Options to Next Steps.* Marlborough District Council, 1-13(13). https://www.marlborough.govt.nz/services/recycling-and-resource-recovery/rubbish-and-recycling-projects/grape-marc-grass-and-greenwaste-repurposing-project.

Jossi, F. (2018). *Sustainable: One company's waste is another's opportunity.* https://finance-commerce.com/2018/03/sustainable-one-companys-waste-is-anothers-opportunity

Kashaninejad, M., & Tabil, L. G. (2009). Resistance of bulk pistachio nuts (Ohadi variety) to airflow. *Journal of Food Engineering, 90*(1). https://doi.org/10.1016/j.jfoodeng.2008.06.007

Kobayashi, N., Hamabe, H., Yamaji, S., Suami, A., & Itaya, Y. (2019). Effect of sludge char addition on drying rate and decomposition rate of organic waste during bio-drying. *Journal of Material Cycles and Waste Management.* https://doi.org/10.1007/s10163-019-00847-z

Kraft Heinz. (2022). *Waste Reduction.* https://www.kraftheinzcompany.com/esg/waste-reduction.html

Kumar, N., & Gupta, S. K. (2022). Exploring drying kinetics and fate of nutrients in thermal digestion of solid organic waste. *Science of the Total Environment, 837,* 155804. https://doi.org/10.1016/j.scitotenv.2022.155804

Lamidi, R. O., Jiang, L., Pathare, P. B., Wang, Y. D., & Roskilly, A. P. (2019). Recent advances in sustainable drying of agricultural produce: A review. *Applied Energy, 233–234.* https://doi.org/10.1016/j.apenergy.2018.10.044

Lim, S. (2019). *Rendering – Essential Recycling.* https://www.ofimagazine.com/content-images/news/Rendering_process2.pdf/, 32–35.

Lipan, L., Rusu, B., Sendra, E., Hernández, F., Vázquez-Araújo, L., Vodnar, D. C., & Carbonell-Barrachina, Á. A. (2020). Spray drying and storage of probiotic-enriched almond milk: Probiotic survival and physicochemical properties. *Journal of the Science of Food and Agriculture, 100*(9). https://doi.org/10.1002/jsfa.10409

Lipan, L., Rusu, B., Simon, E. L., Sendra, E., Hernández, F., Vodnar, D. C., Corell, M., & Carbonell-Barrachina, Á. (2021). Chemical and sensorial characterization of spray dried hydroSOStainable almond milk. *Journal of the Science of Food and Agriculture, 101*(4). https://doi.org/10.1002/jsfa.10748

Liu, L., Huang, Y., Cao, J., Liu, C., Dong, L., Xu, L., & Zha, J. (2018). Experimental study of biomass gasification with oxygen-enriched air in fluidized bed gasifier. *Science of The Total Environment, 626.* https://doi.org/10.1016/j.scitotenv.2018.01.016

Lu, J., Zhang, J., Zhu, Z., Zhang, Y., Zhao, Y., Li, R., Watson, J., Li, B., & Liu, Z. (2017). Simultaneous production of biocrude oil and recovery of nutrients and metals from human feces via hydrothermal liquefaction. *Energy Conversion and Management, 134.* https://doi.org/10.1016/j.enconman.2016.12.052

Ma, J., Zhang, L., Mu, L., Zhu, K., & Li, A. (2018). Thermally assisted bio-drying of food waste: Synergistic enhancement and energetic evaluation. *Waste Management, 80,* 327–338. https://doi.org/10.1016/j.wasman.2018.09.023

Maher, M., Taghian Dinani, S., & Shahram, H. (2020). Extraction of phenolic compounds from lemon processing waste using electrohydrodynamic process. *Journal of Food Measurement and Characterization, 14*(2). https://doi.org/10.1007/s11694-019-00323-0

Martin, A. (2007, November 15). Frito-Lays plans environmentally neutral potato chips. *The New York Times.* https://www.nytimes.com/2007/11/15/business/worldbusiness/15iht-plant.1.8347956.html

McCauley, E. (2022, March 17). How banana plant waste is turned into menstrual pads. *Business Insider.* https://www.businessinsider.com/how-banana-plant-waste-is-turned-into-biodegradable-menstrual-pads-2022-3

Merdun, H., & Sezgin, İ. V. (2018). Products distribution of catalytic co-pyrolysis of greenhouse vegetable wastes and coal. *Energy, 162*. https://doi.org/10.1016/j.energy.2018.08.004

Midilli, A. (2001). Determination of pistachio drying behaviour and conditions in a solar drying system. *International Journal of Energy Research, 25*(8). https://doi.org/10.1002/er.715

Mill, R. (2023). *Our Story*. https://www.renewalmill.com/pages/about-us

Milledge, J. J., & Harvey, P. J. (2016). Golden Tides: Problem or golden opportunity? The valorisation of Sargassum from beach inundations. *Journal of Marine Science and Engineering, 4*(3). MDPI AG. https://doi.org/10.3390/jmse4030060

Mohammed, M., Ozbay, I., Karademir, A., & Isleyen, M. (2017). Pre-treatment and utilisation of food waste as energy source by bio-drying process. *Energy Procedia, 128*. https://doi.org/10.1016/j.egypro.2017.09.021

Mohan, A., Shanmugam, S., & Nithyalakshmi, V. (2016). Comparison of the nutritional, physico-chemical and anti-nutrient properties of freeze and hot air dried watermelon (citrullus lanatus) rind. *Biosciences Biotechnology Research Asia, 13*(2), 1113–1119. https://doi.org/10.13005/bbra/2140

Mohan, K., Muralisankar, T., Uthayakumar, V., Chandirasekar, R., Revathi, N., Ramu Ganesan, A., Velmurugan, K., Sathishkumar, P., Jayakumar, R., & Seedevi, P. (2020). Trends in the extraction, purification, characterisation and biological activities of polysaccharides from tropical and sub-tropical fruits – A comprehensive review. *Carbohydrate Polymers, 238*. https://doi.org/10.1016/j.carbpol.2020.116185

Moussaoui, H., Bahammou, Y., Tagnamas, Z., Kouhila, M., Lamharrar, A., & Idlimam, A. (2021). Application of solar drying on the apple peels using an indirect hybrid solar-electrical forced convection dryer. *Renewable Energy, 168*, 131–140. https://doi.org/10.1016/j.renene.2020.12.046

Muller, E. EL, Sheik, A. R., & Wilmes, P. (2014). Lipid-based biofuel production from wastewater. *Current Opinion in Biotechnology, 30*. https://doi.org/10.1016/j.copbio.2014.03.007

Mutlu, Ö. Ç., Büchner, D., Theurich, S., & Zeng, T. (2021). Combined use of solar and biomass energy for sustainable and cost-effective low-temperature drying of food processing residues on industrial-scale. *Energies, 14*(3). https://doi.org/10.3390/en14030561

Nagel, A., Neidhart, S., Anders, T., Elstner, P., Korhummel, S., Sulzer, T., Wulfkühler, S., Winkler, C., Qadri, S., Rentschler, C., Pholpipattanapong, N., Wuthisomboon, J., Endress, H. U., Sruamsiri, P., & Carle, R. (2014). Improved processes for the conversion of mango peel into storable starting material for the recovery of functional co-products. *Industrial Crops and Products, 61*, 92–105. https://doi.org/10.1016/j.indcrop.2014.06.034

Namir, M., Elzahar, K., Ramadan, M. F., & Allaf, K. (2017). Cactus pear peel snacks prepared by instant pressure drop texturing: Effect of process variables on bioactive compounds and functional properties. *Journal of Food Measurement and Characterization, 11*(2), 388–400. https://doi.org/10.1007/s11694-016-9407-z

Nenciu, F., Stanciulescu, I., Vlad, H., Gabur, A., Turcu, O. L., Apostol, T., Vladut, V. N., Cocarta, D. M., & Stan, C. (2022). Decentralized processing performance of fruit and vegetable waste discarded from retail, using an automated thermophilic composting technology. *Sustainability (Switzerland), 14*(5). https://doi.org/10.3390/su14052835

Nestlé. (2015). *The Nestlé Commitment to Reduce Food Loss and Waste*. https://champions123.org/sites/default/files/2020-09/nestle.pdf

Nestlé. (2023). *Reducing Food Waste and Loss*. https://www.nestle.com/csv/impact/environment/waste-and-recovery

Nestlé Malaysia. (2020). *Zero Waste to Landfill*. https://www.nestle.com.my/stories/zero-waste-landfill

NETZRO. (n.d.). *The NETZRO Platform & Network Converts Industrial Food and Beverage Byproducts into New Revenue and Becomes Part of the Circular Food Economy*. Retrieved January 8, 2023, from https://www.netzro.us/projects

NETZRO. (2023). *Netzro*. https://netzro.us/about/

Ngaini, Z., Jamil, N., Wahi, R., Shahrom, F. D., Ahmad, Z. A., & Farooq, S. (2022). Convenient conversion of palm fatty acid distillate to biodiesel via rice husk ash catalyst. *BioEnergy Research*, *15*(2), 1316–1326. https://doi.org/10.1007/s12155-021-10331-y

Nieto-Calvache, J. E., de Escalada Pla, M., & Gerschenson, L. N. (2019). Dietary fibre concentrates produced from papaya by-products for agroindustrial waste valorisation. *International Journal of Food Science and Technology*, *54*(4), 1074–1080. https://doi.org/10.1111/ijfs.13962

Pasten, A., Uribe, E., Stucken, K., Rodríguez, A., & Vega-Gálvez, A. (2019). Influence of drying on the recoverable high-value products from olive (cv. Arbequina) waste cake. *Waste and Biomass Valorization*, *10*(6), 1627–1638. https://doi.org/10.1007/s12649-017-0187-4

Patel, S. (2015). Functional food relevance of whey protein: A review of recent findings and scopes ahead. *Journal of Functional Foods*, *19*, 308–319.

Pattiya, A., & Suttibak, S. (2017). Fast pyrolysis of sugarcane residues in a fluidised bed reactor with a hot vapour filter. *Journal of the Energy Institute*, *90*(1). https://doi.org/10.1016/j.joei.2015.10.001

Pepi, C. (2022). *Experimental study and sustainability assessment for fertilizers recovery from seafood waste*. Universita Politecnica Delle Marche.

Pepsico. (n.d.). *Waste*. Retrieved January 8, 2023, from https://www.pepsico.com/esg-topics-a-z/waste

Pereira, C. G., & Meireles, M. A. A. (2010). Supercritical fluid extraction of bioactive compounds: Fundamentals, applications and economic perspectives. *Food and Bioprocess Technology*, *3*(3). https://doi.org/10.1007/s11947-009-0263-2

Peterson, E. (2019). *Renewal Mill*. https://companyweek.com/article/renewal-mill

Peydayesh, M., Bagnani, M., & Soon, W. L. (2022). Turning food protein waste into sustainable technologies. *Chemical Reviews*. https://doi.org/10.1021/acs.chemrev.2c00236

Pham, N. D., Martens, W., Karim, M. A., & Joardder, M. U. H. (2018). Nutritional quality of heat-sensitive food materials in intermittent microwave convective drying. *Food & Nutrition Research*, *62*(0). https://doi.org/10.29219/fnr.v62.1292

Pham, N. D., Khan, M. I. H., & Karim, M. A. (2020). A mathematical model for predicting the transport process and quality changes during intermittent microwave convective drying. *Food Chemistry*, *325*. https://doi.org/10.1016/j.foodchem.2020.126932

Pisalkar, P. S., Jain, N. K., & Jain, S. K. (2011). Osmo-air drying of aloe vera gel cubes. *Journal of Food Science and Technology*, *48*(2). https://doi.org/10.1007/s13197-010-0121-2

Plazzotta, S., Calligaris, S., & Manzocco, L. (2018). Application of different drying techniques to fresh-cut salad waste to obtain food ingredients rich in antioxidants and with high solvent loading capacity. *LWT - Food Science and Technology*, *89*, 276–283. https://doi.org/10.1016/j.lwt.2017.10.056

Raghavi, L. M., Moses, J. A., & Anandharamakrishnan, C. (2018). Refractance window drying of foods: A review. *Journal of Food Engineering, 222,* 267–275. https://doi.org/10.1016/J.JFOODENG.2017.11.032

Rahman, M. S., Ed. (2020). *Handbook of Food Preservation.* CRC Press. https://doi.org/10.1201/9780429091483

Ramya, V., & Jain, N. K. (2017). A review on osmotic dehydration of fruits and vegetables: An integrated approach. *Journal of Food Process Engineering, 40*(3). https://doi.org/10.1111/jfpe.12440

Renewal Mill. (n.d.). *Let's Save the Planet by Being Ridiculously Creative with Neglected Nutrients.* Retrieved January 8, 2023, from https://www.renewalmill.com/pages/our-process

Rodriguez, A., Sancho, A. M., Barrio, Y., Rosito, P., & Gozzi, M. S. (2019). Combined drying of Nopal pads (Opuntia ficus-indica): Optimization of osmotic dehydration as a pretreatment before hot air drying. *Journal of Food Processing and Preservation, 43*(11). https://doi.org/10.1111/jfpp.14183

Rodríguez García, S. L., & Raghavan, V. (2022). Green extraction techniques from fruit and vegetable waste to obtain bioactive compounds—A review. *Critical Reviews in Food Science and Nutrition, 62*(23). https://doi.org/10.1080/10408398.2021.1901651

Rosli, M. I., Abdul Nasir, A. M., Takriff, M. S., & Chern, L. P. (2018). Simulation of a fluidized bed dryer for the drying of sago waste. *Energies, 11*(9). https://doi.org/10.3390/en11092383

Rostami, M. A., Mirdamadiha, F., & Golshan, A. (2006). Evaluation and comparison of current pistachio dryers in Kerman province. *Acta Horticulturae, 726.* https://doi.org/10.17660/ActaHortic.2006.726.101

Saavedra, J., Córdova, A., Navarro, R., Díaz-Calderón, P., Fuentealba, C., Astudillo-Castro, C., Toledo, L., Enrione, J., & Ranilla, L. G. (2017). Industrial avocado waste: Functional compounds preservation by convective drying process. *Journal of Food Engineering, 198,* 81–90.

Sagar, N. A., Pareek, S., Sharma, S., Yahia, E. M., & Lobo, M. G. (2018). Fruit and vegetable waste: Bioactive compounds, their extraction, and possible utilization. *Comprehensive Reviews in Food Science and Food Safety, 17*(3). https://doi.org/10.1111/1541-4337.12330

Salehi, L., & Taghian Dinani, S. (2020). Application of electrohydrodynamic-ultrasonic procedure for extraction of β-carotene from carrot pomace. *Journal of Food Measurement and Characterization, 14*(6). https://doi.org/10.1007/s11694-020-00542-w

Salina Md Salim, N., Kudakasseril Kurian, J., Gariepy, Y., & Raghavan, V. (2016). Application and the techno-economical aspects of integrated microwave drying systems for development of dehydrated food products. *Japan Journal of Food Engineering, 17*(4), 139–146. https://doi.org/10.11301/jsfe.17.139

Santos, D. da C., Queiroz, Alexandre J. de M., Figueirêdo, R. M. F. De, & Oliveira, E. N. A. de. (2014). Solar drying of annatto grains and waste grains flour of annatto. *Bioscience Journal, 30*(2), 436–446.

Scalcon, A., Sone, A. P., Johann, G., Gimenes, M. L., & Vieira, M. G. A. (2018). Shrinkage of digested sludge from gelatin production. *Drying Technology, 36*(13), 1603–1618. https://doi.org/10.1080/07373937.2017.1419479

Serrano-Díaz, J., Sánchez, A. M., Alvarruiz, A., & Alonso, G. L. (2013). Preservation of saffron floral bio-residues by hot air convection. *Food Chemistry, 141*(2), 1536–1543. https://doi.org/10.1016/j.foodchem.2013.04.029

Shahram, H., & Taghian Dinani, S. (2019). Influences of electrohydrodynamic time and voltage on extraction of phenolic compounds from orange pomace. *LWT, 111.* https://doi.org/10.1016/j.lwt.2019.05.002

Shih, Y. T., Wang, W., Hasenbeck, A., Stone, D., & Zhao, Y. (2020). Investigation of physicochemical, nutritional, and sensory qualities of muffins incorporated with dried brewer's spent grain flours as a source of dietary fiber and protein. *Journal of Food Science, 85*(11), 3943–3953. https://doi.org/10.1111/1750-3841.15483

Singh, A. P., Mandal, R., Shojaei, M., Singh, A., Kowalczewski, P. L., Ligaj, M., Pawlicz, J., & Jarzebski, M. (2020). Novel drying methods for sustainable upcycling of brewers' spent grains as a plant protein source. *Sustainability (Switzerland), 12*(9). https://doi.org/10.3390/su12093660

Singh, D. B., & Kingsly, A. R. P. (2008). Effect of convective drying on qualilty of anardana. *Indian Journal of Horticulture, 65,* 413–416.

Sinrod, A. J. G., Avena-Bustillos, R. J., Olson, D. A., Crawford, L. M., Wang, S. C., & McHugh, T. H. (2019). Phenolics and antioxidant capacity of pitted olive pomace affected by three drying technologies. *Journal of Food Science, 84*(3), 412–420. https://doi.org/10.1111/1750-3841.14447

Slopiecka, K., Liberti, F., Massoli, S., Bartocci, P., & Fantozzi, F. (2022). Chemical and physical characterization of food waste to improve its use in anaerobic digestion plants. *Energy Nexus, 5,* 100049. https://doi.org/10.1016/j.nexus.2022.100049

Sogi, D. S., Siddiq, M., Greiby, I., & Dolan, K. D. (2013). Total phenolics, antioxidant activity, and functional properties of "Tommy Atkins" mango peel and kernel as affected by drying methods. *Food Chemistry, 141*(3), 2649–2655. https://doi.org/10.1016/j.foodchem.2013.05.053

Song, D. B., Lim, K. H., Jung, D. H., & Yoon, J. H. (2020). Analysis of drying characteristics and cost of high-capacity vacuum-drying food waste disposal system using steam. *Journal of Biosystems Engineering, 45*(3), 126–132. https://doi.org/10.1007/s42853-020-00052-z

Sridhar, K., & Charles, A. L. (2020). Mathematical modeling and effect of drying temperature on physicochemical properties of new commercial grape "Kyoho" seeds. *Journal of Food Process Engineering, 43*(3). https://doi.org/10.1111/jfpe.13203

Suntory. (n.d.). *Promoting the reduction and recycling of waste.* Retrieved January 8, 2023, from https://www.suntory.com/csr/activity/environment/recycle/waste/

Szychowski, P. J., Lech, K., Sendra-Nadal, E., Hernández, F., Figiel, A., Wojdyło, A., & Carbonell-Barrachina, Á. A. (2018). Kinetics, biocompounds, antioxidant activity, and sensory attributes of quinces as affected by drying method. *Food Chemistry, 255.* https://doi.org/10.1016/j.foodchem.2018.02.075

Tahvildari, K., Anaraki, Y. N., Fazaeli, R., Mirpanji, S., & Delrish, E. (2015). The study of CaO and MgO heterogenic nano-catalyst coupling on transesterification reaction efficacy in the production of biodiesel from recycled cooking oil. *Journal of Environmental Health Science and Engineering, 13*(1). https://doi.org/10.1186/s40201-015-0226-7

Tiwari, S., Yawale, P., & Upadhyay, N. (2022). Carotenoids: Extraction strategies and potential applications for valorization of under-utilized waste biomass. *Food Bioscience, 48,* 101812. https://doi.org/10.1016/j.fbio.2022.101812

Turner, T., Wheeler, R., & Oliver, I. W. (2022). Evaluating land application of pulp and paper mill sludge: A review. *Journal of Environmental Management, 317,* 115439.

Tyson. (2022). *Reducing waste is good for our business and the environment as it allows us to streamline costs and send as few materials to landfills as possible.* https://www.tysonsustainability.com/natural-resources/material-resources

Uribe, E., Lemus-Mondaca, R., Vega-Gálvez, A., López, L. A., Pereira, K., López, J., Ah-Hen, K., & di Scala, K. (2013). Quality characterization of waste olive cake during hot air drying: Nutritional aspects and antioxidant activity. *Food and Bioprocess Technology*, 6(5), 1207–1217. https://doi.org/10.1007/s11947-012-0802-0

Uwineza, P. A., & Waśkiewicz, A. (2020). Recent advances in supercritical fluid extraction of natural bioactive compounds from natural plant materials. *Molecules*, 25(-17). https://doi.org/10.3390/molecules25173847

Valadez-Carmona, L., Plazola-Jacinto, C. P., Hernández-Ortega, M., Hernández-Navarro, M. D., Villarreal, F., Necoechea-Mondragón, H., Ortiz-Moreno, A., & Ceballos-Reyes, G. (2017). Effects of microwaves, hot air and freeze-drying on the phenolic compounds, antioxidant capacity, enzyme activity and microstructure of cacao pod husks (Theobroma cacao L.). *Innovative Food Science and Emerging Technologies*, 41, 378–386. https://doi.org/10.1016/j.ifset.2017.04.012

Venugopal, V. (2021). Valorization of seafood processing discards: Bioconversion and bio-refinery approaches. *Frontiers in Sustainable Food Systems*, 5(June), 1–21. https://doi.org/10.3389/fsufs.2021.611835

Wallin, E., Fornell, R., Räftegård, O., Walfridson, T., & Benson, J. (2020). Design and integration of heat recovery in combination with solar and biomass-based heating in a drying plant. *Chemical Engineering Transactions*, 81, 1387–1392. https://doi.org/10.3303/CET2081232

Walter, W. R., Lee D. P., Harold, R. & Scott Butner, R. (1989). *Significance of Food Processing By-Products as Contributors to Animal Feeds: Phase I - Food Processing Survey*. U.S. Environmental Protection Agency. Contract Number: 68-02-04263. 1–27.

Woodard & Curran, Inc. (2006). 10-Wastes from Industries (Case Studies). In Woodard & Curran Inc.(Ed.), *Industrial Waste Treatment Handbook (Second Edition)* (pp. 409–496). Butterworth-Heinemann. https://doi.org/10.1016/B978-075067963-3/50012-6

Yan, L.-G., He, L., & Xi, J. (2017). High intensity pulsed electric field as an innovative technique for extraction of bioactive compounds—A review. *Critical Reviews in Food Science and Nutrition*, 57(13). https://doi.org/10.1080/10408398.2015.1077193

Yi, J., Li, X., He, J., & Duan, X. (2020). Drying efficiency and product quality of biomass drying: A review. *Drying Technology*, 38(15). https://doi.org/10.1080/07373937.2019.1628772

Zanon Costa, C., Falabella Sousa-Aguiar, E., Peixoto Gimenes Couto, M. A., & Souza de Carvalho Filho, J. F. (2020). Hydrothermal treatment of vegetable oils and fats aiming at yielding hydrocarbons: A review. *Catalysts*, 10(8). https://doi.org/10.3390/catal10080843

Index

Note: **Bold** page numbers refer to tables and *italic* page numbers refer to figures.

For Product Safety Concerns and Information please contact our EU
representative GPSR@taylorandfrancis.com
Taylor & Francis Verlag GmbH, Kaufingerstraße 24, 80331 München, Germany

www.ingramcontent.com/pod-product-compliance
Lightning Source LLC
Chambersburg PA
CBHW070726220326
41598CB00024BA/3313